ESSENTIA

how the brain works

JOHN McCRONE

SERIES EDITOR JOHN GRIBBIN

LONDON, NEW YORK, MUNICH,
MELBOURNE, AND DELHI

senior editor Peter Frances
senior art editor Vanessa Hamilton
US editor Eileen Nester Ramchandran
DTP designer Rajen Shah
picture researcher Sarah Duncan
illustrator Richard Tibbitts

category publisher Jonathan Metcalf
managing art editor Phil Ormerod

Produced for Dorling Kindersley Limited by
Design Revolution Limited, Queens Park Villa,
30 West Drive, Brighton, East Sussex BN2 2GE
editor Liz Wyse
designer Lindsey Johns

First American Edition, 2002

02 03 04 05 10 9 8 7 6 5 4 3 2 1

Published in the United States by
DK Publishing, Inc.
95 Madison Avenue
New York, NY 10016

Published in Great Britain by Dorling Kindersley Limited.

A Cataloging in Publication record is available from the Library of Congress.

ISBN 0-7894-8420-X

Color reproduction by Mullis Morgan, UK
Printed in Italy by Graphicom

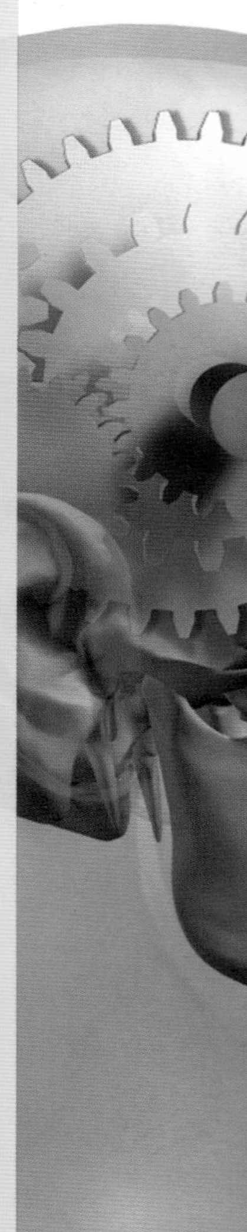

contents

from humble beginnings

The human brain seems impossibly complex, with its billions of cells and trillions of wiring connections. A neuroscientist needs years just to gain a nodding familiarity with the thousand or so major processing regions. Yet, behind all the baffling neural machinery is a deep simplicity of purpose. The brain exists to sense the world in terms of possible responses. Consciousness is often treated as a passive exercise. But to be aware is to be oriented – to have a brain pregnant with goals and expectations. It is the ability to take a selective view of the world. How the brain works is thus the story of how it orientes itself to each passing moment. It turns out that the brain's operating principles are straightforward – so much so that we can see exactly the same principles at work in the most rudimentary lifeforms, such as the humble bacterium. But before we begin our investigation, let us meet the complex lump of matter that every one of us carries around inside our heads.

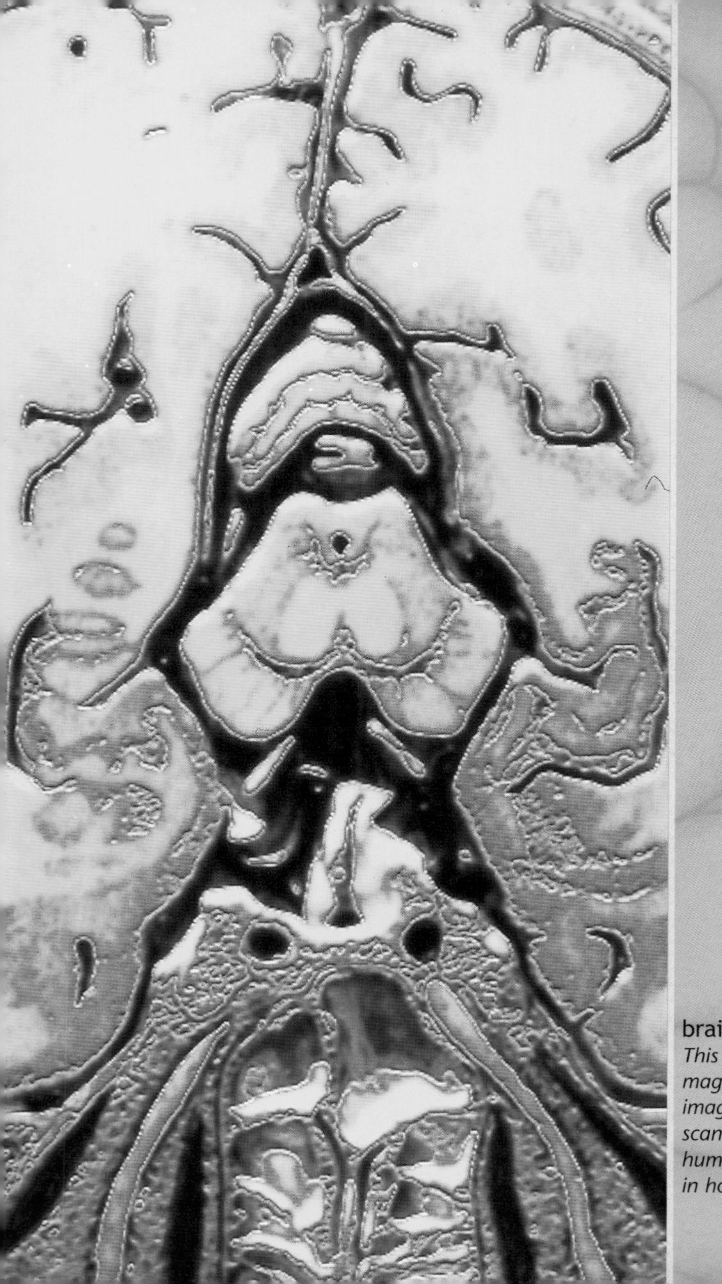

brain scan

This false-color magnetic resonance imaging (MRI) scan of a healthy human brain is seen in horizontal view.

meet your brain

Let's take a look at your brain. Just lift it out of your head and place it on the table in front of you. Well, it is probably somewhat heavier than you were expecting. Its 3lb (1.4 kilograms) would make a big bag of apples. It also looks surprisingly small – about the size of your two fists pressed together. So there it sits on the table, a pinkish fudge color (not gray at all, that is just the color it goes when pickled in preservative) and wrinkled as a walnut. Already it is beginning to slump like jelly under its own weight and ooze clear fluid. Not a pretty sight!

Yet the thing you are now looking at is the single most complex object in the known universe. Packed into that mute lump is a greater density of orderly design than you can find anywhere else.

in the balance
The brain seems to be a solid, heavy lump. But what is it made of? Chemical analysis reveals that it is 78 percent water, ten percent fat, eight percent protein, one percent carbohydrate, one percent salt, and two percent a mix of other minor constituents.

a complex design

A human brain has about 100 billion neurons – that's individual brain cells. And each of these neurons can make anything from a thousand to several hundred thousand synaptic connections – a synapse being the junction between two neurons. So your brain has about 1,000 trillion connections in all. Every one of these connections is meaningful – they are not random bits of living matter. Each connection has its own history, its own wired-in purpose.

This is only scratching the surface of the brain's complexity. Synaptic connections come in many forms, using different messenger molecules to evoke different responses. And neurons themselves make up only a fraction of the brain. The brain also contains glial cells – ordinary cells that carry out support, transport, growth, and housekeeping jobs. There are about 50 of these glial cells to every neuron. Then, nearly half the brain is made up of white matter – the fat-insulated trunk cabling that is used to carry signals about the brain. If the white matter from a single human brain were unraveled, it would form a strand long enough to wrap twice around the globe.

So just imagine... All this – the neurons, their connections, the support cells, the cabling – is balled up inside your skull. And when it is switched on and conscious, the gelatinous circuitry shivers with a traffic of thoughts, impressions, urges, conflicts, worries, curiosities, and intentions.

white matter

Our brain has enough wiring – the trunk cabling that carries signals – to go twice around the world. Some of the connections may span only a few millimeters, but when laid end to end the result is truly impressive.

understanding consciousness

Actually, we have already made a crucial mistake in talking about the brain. While it is usual to think of the brain as the wrinkled lump nestled in your cranium, in fact it penetrates your whole body. The brain itself extends to the tail of the spinal cord. This is gray matter with complicated meshes of neurons, not simply a tract of nerve fibres. Then, from the spinal cord sprouts a maze of nerves that reaches even the remotest corners of the body, catching it in a knowing embrace. Everything from the beating of the heart, the pulsing of the gut, the production of new blood cells, right down to the raising of individual hairs on our arm when we get a fright, all is controlled by the nervous system, and so ultimately the brain. And even where the nervous system does not act directly, it can control by secreting hormone messengers

controlling systems

The brain and spinal cord make up the central nervous system (CNS), which "speaks" to the body through the peripheral nervous system. No cell in the body is isolated. So biologists may think the body is run by its genes, but neurologists know that it is the brain that drives the body through its fine tendrils of information. The mind is a neural empire, in which the conscious self is just the tip of a very large iceberg.

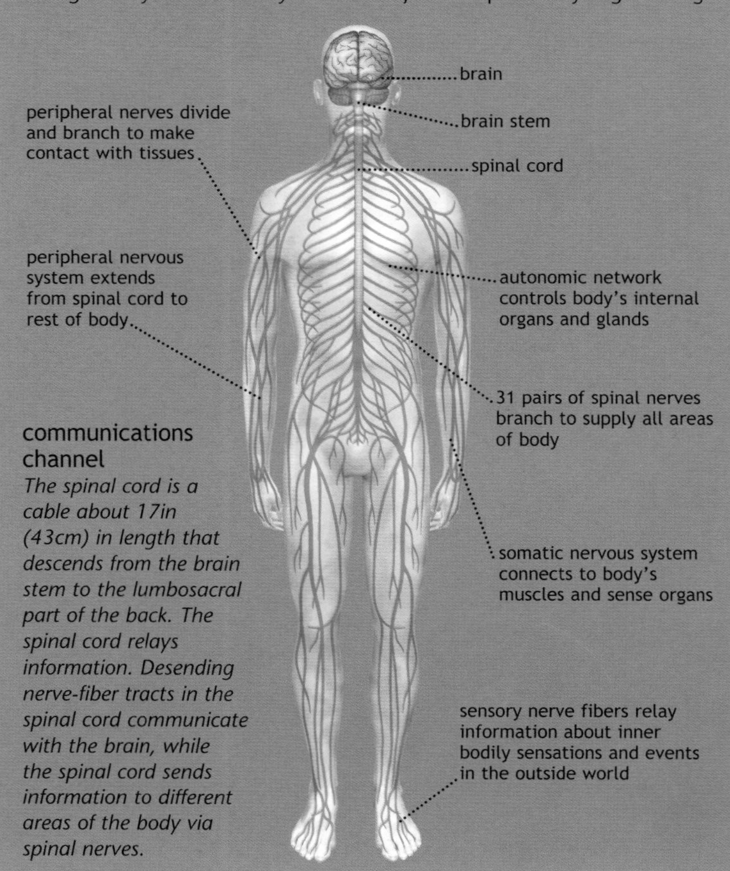

communications channel

The spinal cord is a cable about 17in (43cm) in length that descends from the brain stem to the lumbosacral part of the back. The spinal cord relays information. Desending nerve-fiber tracts in the spinal cord communicate with the brain, while the spinal cord sends information to different areas of the body via spinal nerves.

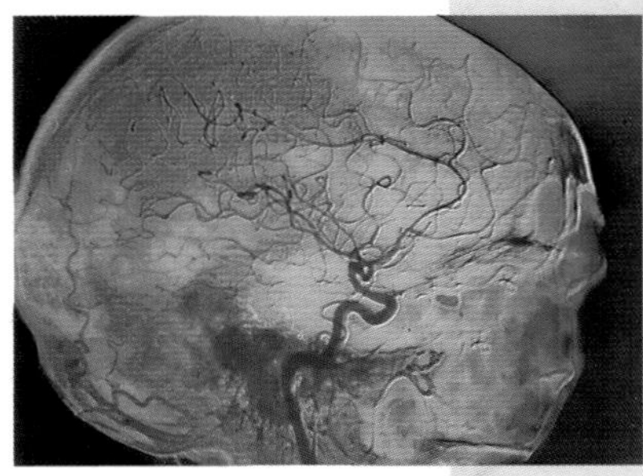

that will diffuse through the blood and body tissues. No organ or individual cell is beyond its reach.

So, if you took your brain out of your skull, it would come trailing a six-foot (2m) long tangle of uprooted fibres and dribbling nerve endings. If consciousness is the product of "brain activity," then it is about more than what takes place inside your head. As psychologists like to say, the mind is embodied. Your whole body sings with knowingness, or "cognition."

an exciting time

So is there a simple story behind the brain's massive complexity? Is there something about the brain that you could learn from a book as apparently slight as the one you are reading?

It happens to be a good time to be talking about brains. The field of neuroscience has undergone an explosion in terms of research. There has been a flood of new techniques, such as brain scans that can take snapshots of brain bloodflow and microscopic glass tubes that can inject just a few molecules of a test drug directly into a neuron. All of these innovations have helped to reveal the organization of the brain in glorious detail. Neuroscientists are no longer afraid to talk openly about consciousness – how the brain produces the mind – which, after all, is what we really want to know.

Although the picture has recently become much clearer, it does not mean that understanding the brain is a particularly easy or familiar task. So now it is time to grab your brain and jam it back into place because it has some serious thinking to do.

energy consumption
An arteriogram shows the network of arteries (red) that supply blood to the brain. The brain accounts for only one-fiftieth of our body weight, yet it demands a fifth of our blood supply.

key points

- The average brain weighs 3lb (1.4kg).
- There are about 100 billion neurons in the brain and about 1,000 trillion neural connections.
- The white matter from a single human brain would wrap twice around the globe.

a living machine

Throughout history, investigators of the human brain have been caught between two styles of explanation – the reductionist and the holistic. Is the brain like a machine in being merely an assemblage of parts, and so a device to be explained in terms of its many parts? Or should we be looking at the brain as a coherent whole, a general fretwork of connections that shapes mental patterns in the same way that the banks of a river shape the tumbling eddies of the current?

gestalt images
The Gestalt psychologists stress that we see the whole before the parts. This is illustrated by common visual illusions. Try as we might, it is hard to combat the holistic way our brains want to interpret the images.

Every era has had its champions of these two opposing views. In the 1930s, for example, the Behaviorists treated the mind as a giant collection of reflexes. They considered thought as just a knee-jerk sequence of associative links. Think of the color red and you may then reflexively think of fire engines or stop signs. Opposing the Behaviorists were the Gestalt psychologists, who believed the brain thinks in wholes before extracting the parts. So the brain seizes on a general thought, forming a broad sense of which direction to head, before doing more work to bring this thought to a specific focus.

“But though all our knowledge begins with experience, it does not follow that it all arises out of experience.”

Immanuel Kant, German philosopher (1787)

Exactly the same argument was taking place a few hundred years earlier. The English philosopher John Locke saw the mind as just an accumulation of sensory and

The English philosopher **John Locke** (1632–1704) was the father of modern psychology. A champion of liberal political reform, he was drawn to a rational view of the mind, arguing that we are born a blank slate, a *tabula rasa*, then are molded by our experiences. We have to acquire habits of perception, networks of mental association, and even the skills of self-awareness. Consciousness is actually built up element by element.

memory fragments, while the German philosopher Immanuel Kant riposted that the mind imposes order holistically, molding thoughts and perceptions to fit its expectations. And nothing has changed today, even though we have learned so much more about the detailed workings of the brain. Neuroscience still has its lumpers and its splitters, those who approach the problem holistically from the top down and those who prefer to build their explanations in reductionist fashion from the bottom up. However, despite this seesawing within science, the view that has filtered through to the general public is by now quite definitely the reductionist viewpoint. Probably because of our great familiarity with machines, technology, and computers – human-made objects that are simply collections of parts – it has become natural to think of complex objects in

holism versus reductionism

Holism is often dismissed as being fashionable but unscientific. Reductionism is seen as "proper" science. But holism simply reminds us that the whole can be more than the sum of its parts. Just as the wetness of water cannot be found within molecules of H_2O, complex systems such as the brain may have emergent properties that cannot be found in an analysis of their components.

the same way. So we are perfectly comfortable when the brain is described in terms of programs, codes, memory stores, and input and output devices – all the standard components of any information-processing system.

not how, but why?

machine mind
It seems easiest just to treat the brain as an elaborate machine – until it is time to put all the parts together.

Certainly, thinking about the brain as a giant computer explains something about the way brains work. But in this book we will try to understand the brain as an organic whole. It is a biological system, and biology does seem to have its own special logic that is not a machine logic.

If we were taking a reductionist approach to the brain, we would now want to identify the basic components of this elegant piece of machinery. We would perhaps talk a bit about the biological components – neurons and their connections. Then we would list the essential psychological components, such as sensation, memory, emotion, cognition, and motor control.

The modular approach of a reductionist strategy says: take an adult human brain and then ignore its history. Don't talk about the brain's purposes – what it is meant to be doing – or how it got to be the way it is. Instead, just

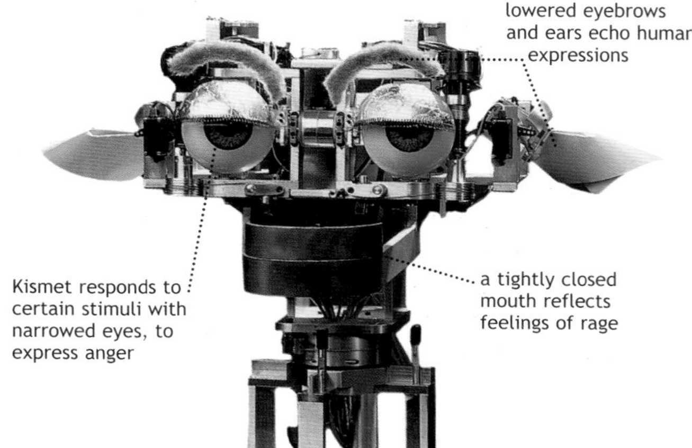

building a mind
This is Kismet, which was built by the artificial intelligence lab at the Massachussetts Institute of Technology. It is designed to have the expressive face of a human, and learns through social interaction, responding to stimuli with emotions. But despite 40 years of such research, consciousness in a machine seems as far away as ever.

examine the anatomical structure of the finished object. This is what will constitute the explanation. When we want to understand how a clock or a car works, we don't need to know who made these pieces of machinery, or even why. We just need a design blueprint. What makes a clock tick or a car move is self-evident from a knowledge of its components and how these connect to each other. The "how?" is visible in the detail of the structure.

The holistic approach starts at the other end of the scale, even before any physical structure exists, and asks the question "why?" The assumption is that every biological organ has evolved to fulfil some purpose. There is something that a brain wants to be, and that desire will inform its design. So the "how?" becomes a secondary question. Nature starts with a "why?" and then shapes itself to do the "how?" First the need and then the structure.

This probably sounds a quibbling point. Of course brains evolved, you might say, just tell me how they work. Again, the holistic view is that the why drives the how. Raw protoplasm could have been turned into just about anything by the crafting forces of evolution. So, to make sense of the anatomy of a brain, you need to have a clear understanding of what it was that evolution was pressing to achieve. The "why?" needs to be made visible in the brain's structure. The brain is, after all, a purposeful organ and its story needs to be told in purposeful terms.

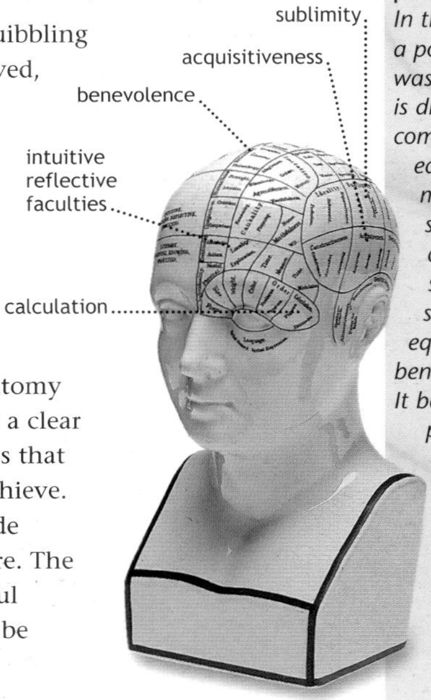

phrenology
In the 19th century, a popular idea was that the brain is divided into compartments, each dealing with mental functions such as acquisitiveness, secretiveness, self-esteem, equanimity, and benevolence. It became a Victorian parlor game to feel the bumps and hollows of a friend's skull, looking for a large bulge of benevolence or a missing zone of sublimity.

views of the brain

Although the brain is not a machine, it does have an anatomy. Areas of the brain are specialized for various activities such as planning movement, making social judgements, or mapping the visual scene. However, any mental action requires the coordinated action of many regions. These diagrams can only hint at the true anatomical complexity of the brain. The cerebral hemispheres are the wrinkled mass of the higher brain. Their outer surface – the cortex – has several hundred distinct processing zones alone. Tucked beneath its folds are the many hundreds of nuclei, glands, organs, and other structures that make up the lower brain.

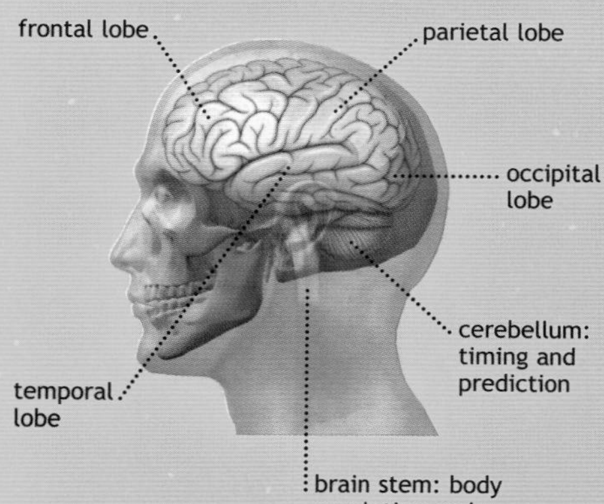

anatomy of the brain

The lobes are the brain's major physical divisions. The frontal lobe handles conscious planning and motor control. The temporal lobe has key memory and auditory centres. The parietal lobe deals with bodily and spatial senses. The occipital lobe handles vision.

scans of the brain

In the 1990s, scientists developed a whole range of techniques to visualize the brain and its activity. MRI uses powerful electromagnetic and radio waves to create cross-sections of the physical structure of the body. PET scans display brain activity. In the PET scan shown here, yellow indicates high activity and blue indicates low. CT uses fine X-ray beams to create images that reveal tumors and blood clots.

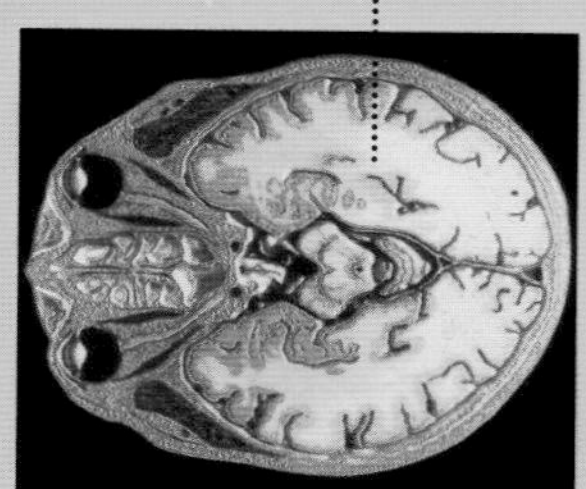

MRI (magnetic resonance imaging) scans reveal the physical structure of the brain

locating the functions

Particular jobs are dealt with by specific parts of the brain. But really the brain is a seamless network of connections; these regions are just the focus for various contributions to the overall flow of neural activity.

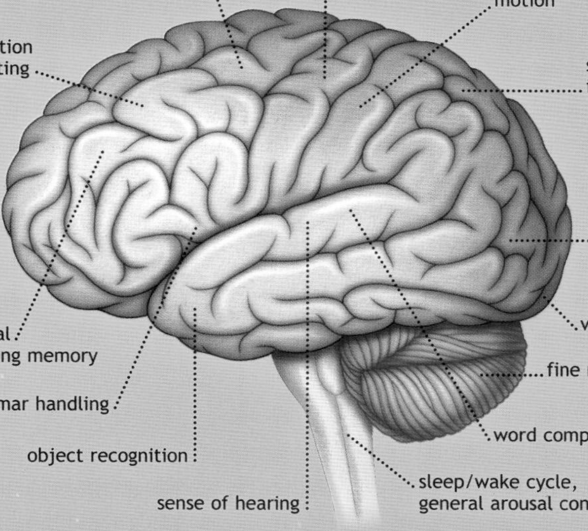

muscle commands
motor planning
touch and motion
attention directing
spatial intelligence
visual motion
vision
spatial working memory
fine motor control
grammar handling
word comprehension
object recognition
sleep/wake cycle, general arousal control
sense of hearing

area of high brain activity

area of low brain activity

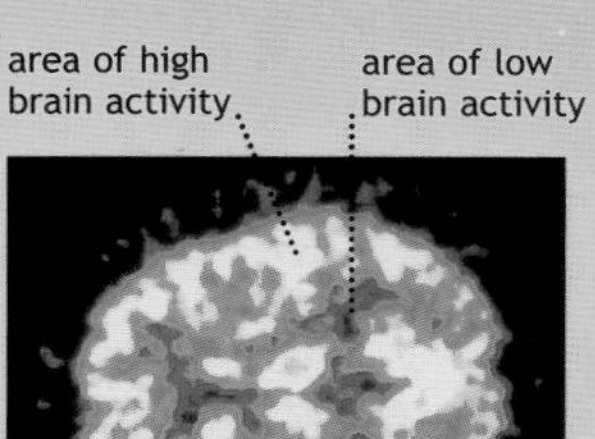

PET (positron emission tomography) scans reveal areas of high and low brain activity

longitudinal fissure separating cerebral hemispheres

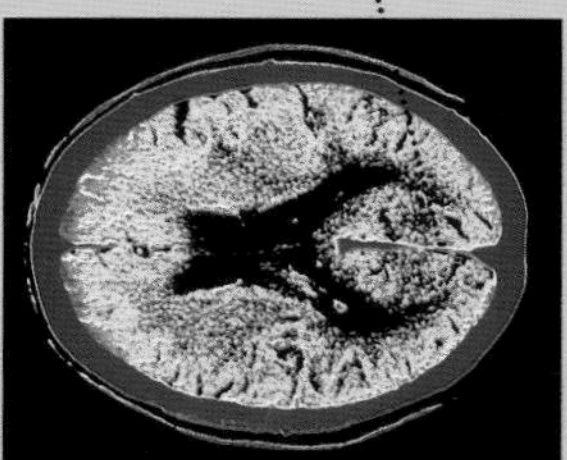

CT (colored computed tomography) images display the density of the brain tissues

why do you need a brain?

Life on Earth began some three and a half billion years ago with simple algae, bacteria, and other single-celled organisms. These primitive creatures had no nervous systems, but they did need to know something about their worlds and then how to respond accordingly.

stop and go

The E. coli *bacterium that lives in our guts has a clever swimming mechanism. When the hairlike flagella on its surface spin one way, they tangle up to form a screw propellor. Spun the other way, they untangle, tumbling the bug until it finds a new direction in which to head.*

A bacterium – for example, *Escherichia coli*, which colonizes the human gut in great numbers – swims along, driven by the anti-clockwise corkscrew paddling of its hair-like flagella, on the lookout for food. Lining its surface are receptors, large protein structures that can sense sugars, such as galactose, or amino acids, such as aspartate. The receptors do this by having binding sites that physically lock onto the target molecules. When they snare a sample, this causes them to change shape, triggering a cascade of chemical signals that run back though the bacterium to the flagella.

While the receptors are picking up traces of food, the instruction to the flagella is to keep rotating anti-clockwise. But if the concentrations decline, then the flagella are told to reverse rotation. The bacterium is sent into a tumbling spin until the scent begins to pick up again, at which point straight-line

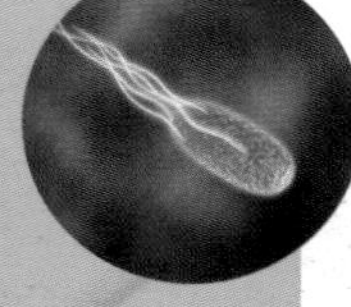

swim

swim

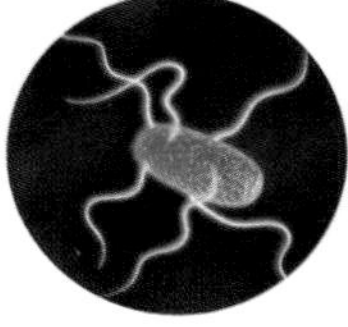

tumble

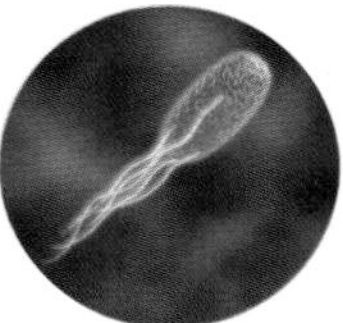

swim

swimming toward the source of the food resumes. The receptors can also lock onto substances the bacterium wants to avoid, such as fatty acids, triggering the opposite response. Meeting increasing concentrations of a fatty acid causes a bacterium to tumble in order to escape.

Of course, to be able to sense changing concentrations of either a repellant or attractant, the receptor proteins need a memory. Binding causes alterations in their internal shape so that for some seconds afterward they become more excited by further traces of a molecule. Their signals to the flagella become more enthusiastic. This way, the bacterium can tell whether it is moving toward or away from concentrations of food or toxins.

Even our immune system has a form of cognition, argued Chilean biologist **Francisco Varela** (1946–2001). Varela used the term autopoiesis – self-production – to describe the way natural systems need to be self-knowing and self-making. The immune system, for example, is a network pervading the body that recognizes what is self and removes proteins or cells that should not be there. Nature builds surprising intelligence both into the bugs that inhabit us and the defenses we use against them.

optimizing behavior

The response path of *Escherichia coli* is a form of brain, and here we have found the reason for any kind of brain, any level of sentience. Brains exist to optimize behavior.

Consciousness is not passive contemplation. It is about seeing through to the particular aspects of the world that demand a response. Brains exist to turn inputs into outputs as quickly and effectively as possible. So brains are for action. And this is the purpose we will find written into their anatomy.

smart bugs *Creatures without brains, such as these* Staphylococcus *bacteria, employ a remarkable variety of senses. They have detectors for oxygen, light, and even magnetic fields. Food-sensing receptors can be grouped at one end of a cell, working in combination to form a discerning "nose."*

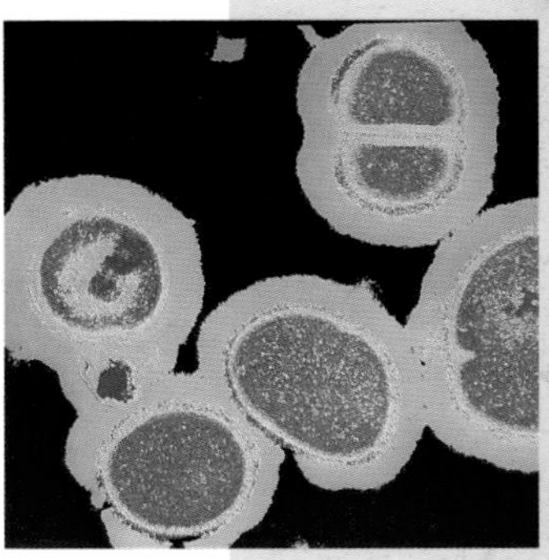

three degrees of adaptation

Brains are shaped by adaptation on at least three time scales. First, there are the adaptive changes that take place on a genetic time scale. Encoded in DNA is the basic blueprint of an organism's brain and nervous system. This blueprint is like a frozen set of good ideas and sensible expectations. It builds a network of connections that is geared in a general way to the business of eating, resting, reproducing, and surviving.

fish out of water
If a biologist from an alien planet were confronted with a single specimen of an Earth fish, what could he predict about the life this organism must live? The genes, and the adaptations they produce – such as fins, gills, and scales – are a kind of cognition, or knowledge, that reflects the kind of environment in which the organism

Even the kinds of sensory organs and output structures with which an organism is born are a type of genetically encoded knowledge. A hoof, hand, or flipper, for example, is a concrete prediction about the kind of world that an organism is going to have to survive in. The same is true of a nose tuned to certain scents or an ear tuned to certain frequencies.

precision response

Once an individual being is born, it starts to learn for itself. The genetic legacy becomes fine-tuned by a personal history of experiences. The individual becomes adapted to its own particular world. Its response pathways come to reflect a private story of the threats and opportunities that it has learned to deal with.

Then, finally, there is the adapting that the brain does in the state of its circuitry in the course of a single moment. Every instant presents a novel set of challenges, and the nervous system must mount an equally precise mental response. The general understandings and general reactions built in by the genes and a lifetime of experiences must be tightened to form the particular pattern of linkage that best handles the current moment. It is as if the brain is zeroing in on the meaning of each moment, following an ever-narrowing cone of adaptive activity.

This is certainly what happens with the humble bacteria wandering our guts. Millions of years of evolution have equipped them with a basic machinery of chemical reactions and tracking behavior. During the lifetime of each individual bacterium, the settings of this inherited machinery must be adapted to the particular environment. Then, from moment to moment, the bacterium's chemo-reception pathway is alive with knowledge of whether the concentrations of food are growing stronger or weaker, or whether it is time to dive for cover as a burst of acid indigestion passes by.

By adapting across multiple time scales, the bacterium homes in on what is going on its world. And, as we shall see, exactly the same is true of human brains – although this is a tale with perhaps a surprising twist at the end.

under the microscope

Bacteria are responsive to what is going on about them from moment to moment. Scrutiny of the human brain shows that it adapts in much the same way. All that changes is the scale, not the principles involved.

switching on and off

Brains are for action. But also inaction. It is just as important for an animal to know when it can relax, chill out, and save energy. So, in assessing each moment, the brain is continually making decisions about whether to raise or lower the body's general state of arousal. Heart rate, blood pressure, breathing, sweating, even the state of your digestion will be tuned to suit the conditions of the moment.

creating consciousness

The purpose of a brain is to optimize behavior – to juggle the body's needs against the threats and possibilities of the moment. Somehow in doing this, the brain generates awareness. Organized states of neural activity are experienced as organized states of mental activity. It is hard to explain the workings of the brain in terms that close the gap between the objective world of neural circuits and the subjective world of the mind. But then you probably don't realize that subjectively your consciousness must lag behind the state of the world, or that much of your brain activity is short-circuited through the use of subconscious habits, or that every moment begins with a sharply defined set of mental expectations. Objective descriptions of the processing mechanisms of the brain may lead you to some surprising realizations about what actually goes on inside the very private world of your own head. Conciousness seems instant and effortless. But in fact there is frantic activity behind every instant of awareness.

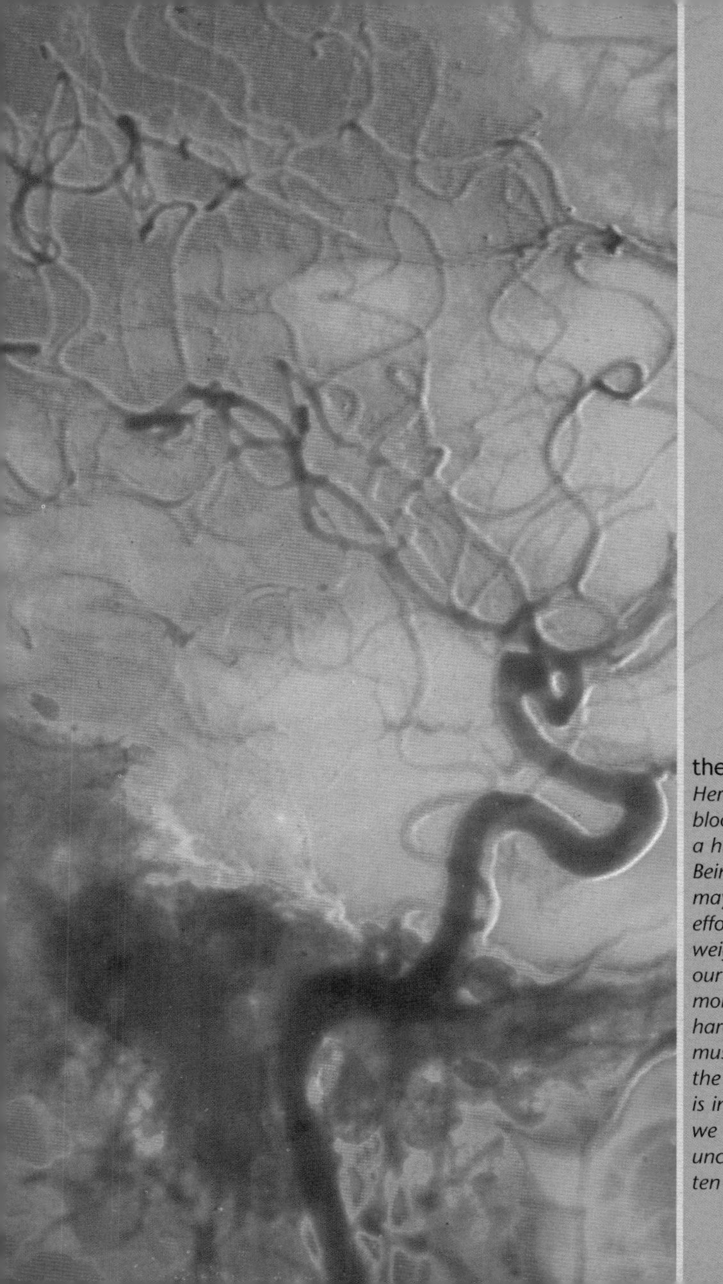

the hungry brain

Here, knots of blood vessels feed a hungry brain. Being conscious may seem effortless, but weight for weight our brains burn more energy than hard-working muscle. And if the blood flow is interrupted, we will become unconscious within ten seconds.

neurons and networks

Life depends on information. To organize unruly matter into forms that can tap sunlight and other energy sources takes some level of sentience.

DNA can be thought of as a memory molecule. The genes encode a recipe to make a body. Over billions of years, they learn to make the kinds of bodies that are likely to be successful. Each new generation – assembled with just a smidgen of variation to explore a slightly wider range of evolutionary possibilities – is like a fresh prediction about what should work. This prediction is then tested by Darwinian selection. Out of the heat of competition comes change. The fittest survive, and the genome is left slightly better prepared for its next round of predictions, completing the cycle of adaptation.

memory molecule

A DNA molecule is no more than a code for the chaining together of a sequence of amino acids. But each strand of amino acids then folds itself up into the shape of a protein, the basic building material of every cell.

But DNA is only one of the forms of information that sustain life. Boundaries, such as the membrane of a cell, the skin of a body, or the lining of the gut, are also informational. These active structures need to be able to distinguish between self and non-self. A cell membrane, for example, has receptor-controlled pores and other structures that can recognize which substances to let in and which to push out.

This general dependence on information is important since it shows that consciousness and brains emerged from something. Bodies already needed multiple levels of cognition and control. So the evolution of nervous systems was just a further step. A nerve cell (which we call a neuron in the brain) is simply an exaggerated version of an ordinary cell. All cells have the ability to secrete and respond to chemical messages. All cells also have a natural difference in electrical charge between their inside and outside. The need to maintain a salty interior means that every cell has to be able to pump out positively charged ions such as sodium. This leaves the cell with a slightly negative resting potential. So, the basic machinery of chemical signaling was in place, and cell membranes already had electrical properties. All that remained to make a neuron was to give the cells a shape – to stretch them out to make an input-output pathway.

understanding neurons

At one end, a neuron has a bush of dendrites, the synapse-studded tendrils that receive input signals. At the other end, a neuron is stretched into the long output arm of an axon. Input electrical charges flow up the dendrites and accumulate on the cell body. Once enough charge has gathered to exceed the neuron's threshold, it fires. A spike – a rolling wave of depolarization (a change in electrical potential across the cell membrane) – speeds down the axon toward its destination.

Physically, the difference between neurons and ordinary cells is slight. But in terms of a capacity to represent information, the differences are immense. The shape of an individual neuron, the spatial pattern of connections it makes with other neurons, and the temporal patterns formed by the spikes it then transmits can all carry meaning.

conduction speeds

The brain is like a road network with a few fast highways and a maze of tiny back roads. Main nerve paths conduct at 248 mph (400 km/h) but most traffic crawls at less than 12mph (20 km/h).

how a neuron works
A rolling wave of electrical discharge speeds down the insulated axon to the synaptic junction, triggering the release of chemicals that stimulates the next neuron in the chain.

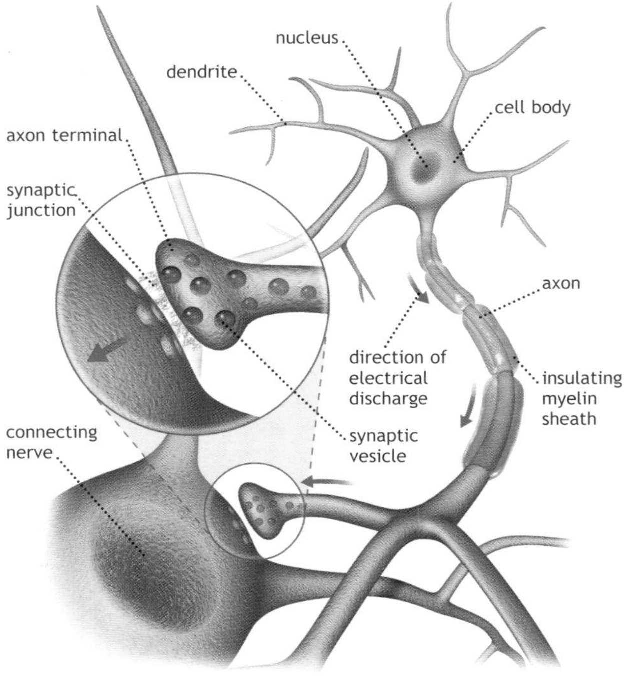

nucleus
dendrite
cell body
axon terminal
synaptic junction
axon
direction of electrical discharge
insulating myelin sheath
connecting nerve
synaptic vesicle

As we have seen, all cells traffic in chemical messages. But the delivery of such messages is somewhat haphazard. The signals have to diffuse through tissues, or be borne along by blood vessels, to reach their targets. Neurons, on the other hand, can deliver a message point-to-point, anywhere in the body, almost instantly. Thickly insulated motor nerves conduct their signals at several hundred miles per

MEG scans
Magnetoencephalography (MEG) is a technique used to measure magnetic fields generated from nerve activity in the brain. This scan shows nerve cells in the motor area of the brain sending commands milliseconds before the patient starts to move a finger.

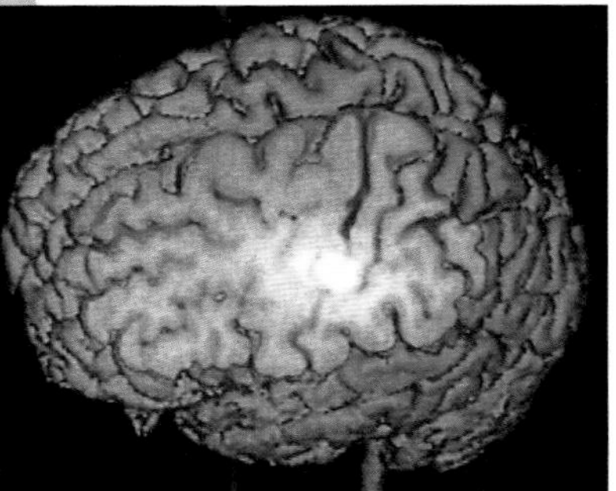

hour. The messages themselves are coded in flexible patterns of spikes. A neuron can be conveying information about sights or sounds, tastes or motions, urges or frustrations. The spikes look just the same; the meaning lies in the fleeting patterns of connection being formed.

A single nerve can act as an intelligent pathway. Wired at one end to pressure sensors, and at the other to muscles, a nerve can do things like allow a worm to recoil when prodded. But a network of nerves all talking to each other is much better, since a network can come to represent complex realizations and behaviors.

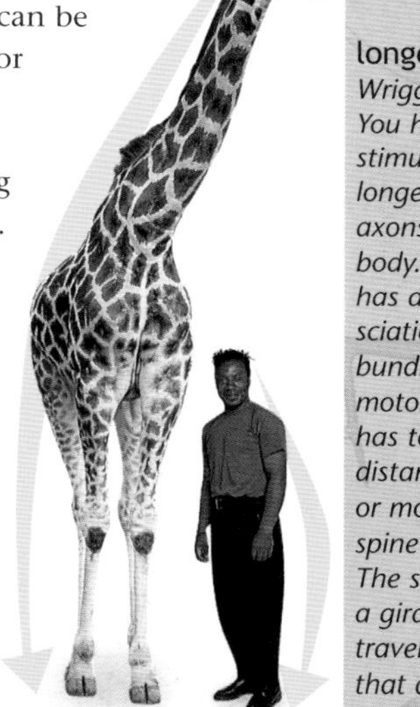

longest axons
Wriggle your toes. You have just stimulated the longest nerve axons in your body. Your brain has activated the sciatic nerve, a bundle of tiny motor fibers, which has to cover a distance of a meter or more from the spine to the foot. The same nerves in a giraffe have to travel five times that distance.

the neural landscape

In brains, every neuron is connected to several thousands of its neighbours. The strength of each of these connections is tuned by experience. The receptors at a synapse may change their shape following a burst of activity. For a second or so, a connection will be made more sensitive. In the longer term, the tip of a dendrite may swell to expose new synapses, physically strengthening a connection, or a neuron might sprout entirely new dendrites. The number of ways of adjusting the connection between two brain cells – of wiring in a memory – runs into the dozens, probably even hundreds.

The result of all this careful tuning is a neural landscape sculpted by its experiences. A network of nerves starts off in a neutral state, but through memory changes it becomes a surface etched with bumps and hollows. When fresh input flows into this landscape, it is then channeled down well-trodden pathways. It is the shape of the network that does the processing, although of course new inputs, new experiences, will in turn start to carve out new memory paths. The network is always learning at the same time as it is making connections.

feedback systems

The nervous system of a worm is small enough for the genes to specify the placement of each individual cell. So the right pattern of connections controlling, for example, the recoil reflex, can be wired in over many generations. Simply put, those worms that know exactly when to pull their heads in are the ones that will survive to pass on their wiring plans.

wiring plans
A worm has a hard-wired nervous system of just a few tens of thousands of neurons. Producing such a nervous system is made all the easier because the body is segmented. Genes simply code for the nerves for one segment, then repeat.

But, as we shall see, the tale of brains has been one of increasing feedback speeds. The cycle of adaptation has spun faster and faster until the connective pattern of brains have come to be reshaped on the fly. In a worm, the processing is certainly extremely rapid – the signal to recoil flashes across its nervous system in an instant. But the learning process is very slow; it takes many generations to restructure the genetically coded pattern of connections.

With large brains, however, both processing and memory changes have become almost instant. So what was originally a neural landscape being slowly sculpted by genetic experience has evolved to become an active process of "in-the-moment" neural resculpting that *causes* experience. The brain's response is no longer merely reflexive, a series of patterns emitted by fixed pathways. The brain now knows why it is acting the way it is acting. It is making choices. It is aware.

anatomy of complex brains

The human brain has billions of neurons and trillions of synaptic connections. Fortunately, it has only a few thousand distinct processing areas – lobes, ganglia, nuclei, and other gobbets of gray matter large enough to have individual names and known jobs. Unfortunately, it takes a neuroscientist years to learn them all. Here we are going to have to content ourselves with the scantiest of sketches. So, to get an idea of the basic plan of the human brain, let us once again consider its evolutionary origins.

Any brain is no more than an elaboration of the nervous system, a complication on the pathway connecting sensation to action. The very simplest nervous systems, such as that of a jellyfish, have no center. They are just a web of nerves that when prodded anywhere will spark a muscular contraction. But in evolutionary terms, it was clear that it was good for an animal to have a definite head and tail – a body plan that expressed an intent, a direction.

"To understand the brain, is important to grasp that is the end product of a ng process of evolution y natural selection."

ancis Crick (1994)

The impact of a linear body plan is dramatic. For a start, being worm-shaped makes it easier to evolve a gut – a digestive tube with an input and an output that "processes" food in a sequence of stages.

Jellyfish have to make do with a simple sac, whereas a worm-shaped creature is also always pointed in some definite direction. It is always coming from somewhere, and going somewhere, which helps makes its "processing"

recap

The **brain** is often talked about as a machine or computer. But it is a biological organ that has to grow and evolve. Its sole purpose is to make smart decisions and it does this by establishing networks of connections – nerve pathways – tuned by both genetic and personal experience.

which way is up?
Jellyfish don't have much brain, but "pacemaker" nerves regularly contract their bell to allow them to swim. Jellyfish can detect light and taste, and their touch-triggered sting is one of the fastest known responses in nature. They must be doing something right to have survived for the past 700 million years.

a jellyfish moves in any direction in response to stimuli

of its environment – its locomotion and exploration – a more orderly and deliberate affair. A restriction on choice also reduces the opportunity for confusion or indecision.

finding a purpose

A worm, maggot, or slug already seems a much more purposeful animal than a jellyfish. With a linear body plan, it is only natural also to place all the major sense organs up front near the mouth-parts, which is where the action is. And then it is only logical to build a brain there as well, elaborating the nervous system at the point where all the information from the various sensory systems can be pooled to form a general picture. It is this general picture that is used to drive the animal's behavior.

one-way traffic
Millipedes have up to a few hundred pairs of legs, or as few as a dozen in some species. But what an advantage to have only one head as they bulldoze through the leaf litter.

The vertebrate brain developed according to a linear plan – the sense of smell, vision, and body control were strung out like beads on a chain, while nerve signals flowed back and forth to knit their activity together. There remained a generally connected, jellyfish logic to the vertebrate brain. Tug any corner of the nervous system and the rest would jangle in a coordinated network of response.

As the vertebrate brain grew larger, simple bulges broke up into hundreds of sub-modules, each sharing aspects of the tasks. So with vision, for example, the brain broke up into areas dealing separately with color, motion, shape,

brain evolution

The primitive vertebrate brain – the ancestral brain of animals that have backbones, such as fish, reptiles, birds, and mammals – developed according to a linear plan. It began as a simple neural tube, a segmented spinal cord, which then swelled to form a series of bulges at one end. A forebrain dealt with the sense of smell and associated behaviors, such as eating and mating. A midbrain then dealt with vision and other distance senses such as hearing. Finally, the bulge of a brain stem (the hindbrain) made general decisions about arousal levels and motor activity, being well-positioned to turn the body's thermostat up or down, depending on the demands of the moment. Each of these primitive bulges was intimately connected – nerve signals flowed back and forth to knit their activity together. As the vertebrate brain grew larger, both the number of divisions and the counter-balancing mechanisms of integration became more complex.

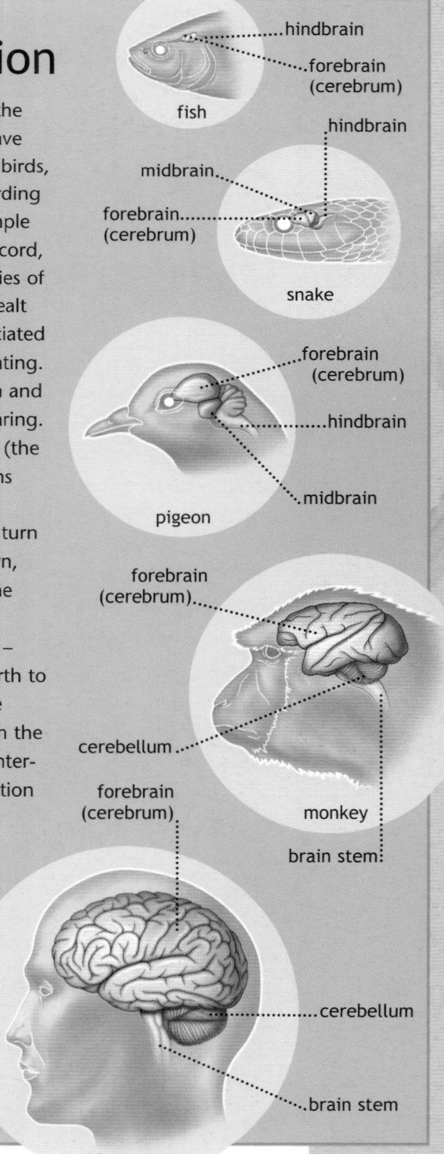

expanding brain

As can be seen, all modern vertebrates share the same basic brain blueprint. The primitive arrangement of a string of bulges at the head of the spinal cord is still clearly visible in fish.

location, and perspective. With more neurons to throw at the job, the processing of visual experience became divided into an ever-increasing number of parts.

Yet to keep all this extra activity integrated, the cross talk between the many brain areas had to increase to match. Hence the great weight of white matter trunk cabling. Neuroscientists estimate that for every one neuron crunching input, another nine are needed to keep the brain informed. So a division of labor and the integration of this labor were paired trends in the elaboration of the vertebrate brain.

the evolution of large brains

Two further trends characterize the evolution of large vertebrate brains. One is encephalization – the ballooning of the very first spinal cord bulge, the forebrain, to create the cerebral hemispheres. Their surface, the cortex, was then colonized by vision, hearing, motor control, and other activities originally located in the lower brain.

The second trend was a change from a hardwired nervous system, built to a genetic template, to a nervous system that was plastic, shaped largely by experience. For once a brain is more than a few million cells in size, the genes can specify only roughly where any block of cells is to be placed. Wiring up the cells must then be carried out through trial and error learning.

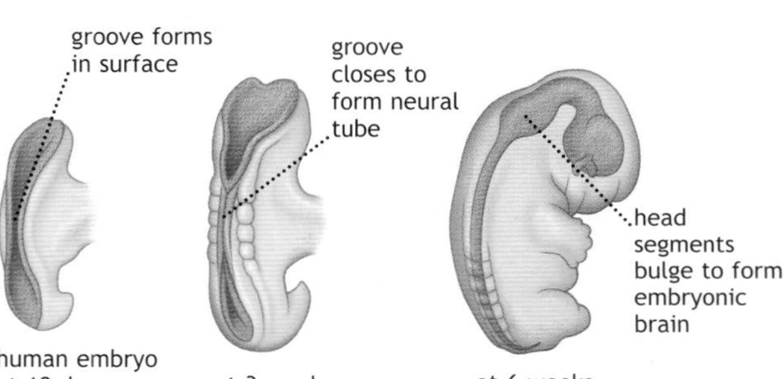

embryo brains

All vertebrate brains form in the same way. For the first few weeks, the embryo is little more than a ball of cells. Then a groove forms in the surface. This sinks until eventually closing over to make a neural tube. The tube becomes segmented and the head segments bulge to become the embryonic brain. The remaining segments form the spinal cord.

tuning into the world

Is a baby conscious when it is born? This will sound a strange question to any parent. But it is a serious one for neuroscientists.

A human baby is born with hardly any connections in its cortex, its higher brain, just a mass of unwired cells. The lower brain is well developed at birth and is capable of producing a variety of instinctive behaviors such as suckling, crying, recoiling, even tracking objects with the eyes. But the higher brain is blank of the memories and experiences with which to make sense of the world.

empty headed?
A newborn baby's brain has cells but not connections. Only the most instinctive parts of the brain are already wired.

a firm foundation

The human brain is a bit like an ice-cream cone, both in its physical shape and its development. There is a firm base. Over millions of years of vertebrate evolution, the genes have wired up the brainstem and many of the mid-brain structures to handle basic jobs such as breathing and swallowing, or even walking, pouncing, and copulating. So the lower brain provides a narrow but solid foundation of the most essential input-output pathways (and for a simple animal, such as a frog or snake, it can handle everything that needs to be handled).

But mammals, and especially the monkeys and apes, have an enormously expanded forebrain. The cerebral hemispheres

cone brain
The brainstem forms a firm but narrow base. The weighty and wrinkled cerebrum awaits the lick of experience.

now sit perched atop the crusty cone of the lower brain like big soft scoops of icecream. They are also produced quite differently. Being much too big to be hardwired, the genes have to throw up a generalized mass of neurons, then allow the experience of life to lick it into shape.

"Children lose on the order of 20 billion synapses per day between early childhood and adolescence. While this may sound harsh, it is generally a very good thing."

Lise Eliot, neurobiologist (2000)

Thus a baby probably enjoys only a reptilian level of awareness – and a somewhat dazed reptile at that! A newborn human certainly has a starter kit of lower-brain instincts and reflexes. But then it has to develop its way to a human level of consciousness by sorting out its higher processing circuits. And given the initial state of these circuits, their earliest inputs are perhaps more confusing than helpful.

constructing a brain

The process of building the higher brain actually starts in the womb. The womb may seem a watery twilight world, but a fetus can still squirm about enough to begin to educate its motor circuits. It also has a chance to touch, taste, and hear. So some learning is possible in the months before birth.

swelling brain

By 14 weeks, a fetus will wriggle when it is touched. When pricked by a needle, it produces stress hormones, suggesting it can feel pain. By 24 weeks, its brain is sufficiently wired up for it to hear, smell, taste, and even blink if bright light is shone onto a mother's stomach.

However, it is only after birth that the higher brain really starts to get going. The cortex neurons enter a phase of rampant growth, sprouting a profusion of dendrites and axons. In the first few years of its life, a baby's brain is forming nearly two million new synaptic connections each second. But this growth is rather random, connections being made willy-nilly. The connections are also immature, not yet having white-

brain cells crawl

To make a brain, neurons have to crawl into position. Cells form in a budding zone around the embryonic neural tube. Then they begin to slither up guidelines laid down by a special type of glial cell. Eventually, chemical markers tell them when they are in the right position. They can stop and start branching, seeking contacts. In this way, a simple neural tube can build itself into a complex three-dimensional mass of neurons.

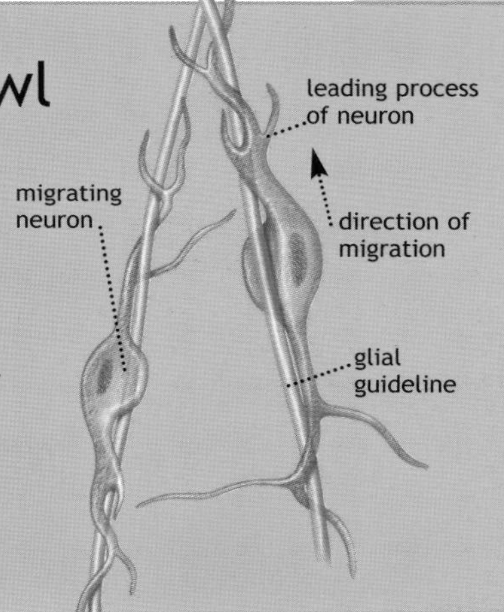

matter sheathing to insulate them. So instead of nerve signals zipping about at hundreds of miles per hour, they are creeping along at nearer walking pace.

By about six months, a baby's brain has made about twice as many connections as it actually needs. The next step is to start pruning these thickets. In fact, the synapses compete against each other to find which is the best placed for the job of processing information about the world. The ones that are in the right position to do useful work will survive. But inactive synapses wither. As a result, neural connections start to disappear at the rate of a quarter million each second. From this pruning, the original mish-mash of wiring is slimmed down to a working set of cortical pathways.

For the baby, the effect must be rather like tuning into a distant radio station. Scientists have made recordings from the brains of infants staring at a pattern of simple black bars. In the

tuning in

As brain connections are pruned, the sensory signals quickly become crisper.

first few weeks of life, the pattern will provoke a broad crackle of activity from many neurons. Because of the promiscuous maze of interconnections along the pathway bringing information from the eyes, plenty of neurons feel they are seeing something. But they cannot be specific. A grid with thicker or thinner lines will evoke much the same general response. Mentally, the baby must experience a rather indistinct state – a sense of vague "grid-ness."

cat or dog?

The difference seems pretty obvious to us. But does the infant eye really see more than a hairy beast with four legs and a tail? Remember that as adults, we can look at a field of sheep and find them indistinguishable. But to the shepherd, tiny differences will be glaringly obvious. Eyes need educating before they can really "see".

A few weeks later, however, the boundary-detecting areas of the visual cortex have begun to tune into the world. The maze of connections has been drastically pruned to give more precise responses. Only particular cells fire to lines of particular size. So now the baby will see thin bars and fat bars as clearly different kinds of experience. Out of a gray crackle of static, the brain gradually homes in on a crisp signal. It tunes into a sharp experience of life.

tuning in

First the infant brain tunes into the simple aspects of the world. It gets used to comprehending shapes, colors, and motions. It learns about simple motor activity, too – how it can use the information about what it senses to control its actions. A baby's first discoveries are about how to grip objects and bring them to its mouth. Then each newly acquired set of sensory and motor skills becomes the platform for yet more complex skills. Once it can see shapes and colors clearly, a child is ready to notice the difference between family and strangers, or cats and dogs.

Steadily, the higher brain bulks up with memories and habits. Its circuits come to represent the astounding variety of things that every human child knows how to do. The problem then is to get it to do the one right thing that best fits each passing moment.

seeing things

Lift your eyes and look around. It is so easy. The world just floods in, an instant panorama. Nothing is missed, nothing is invented, and nothing is incoherent. OK, now forget the all evidence of your eyes, because seeing the world is not like that at all.

How do your brain pathways process what you experience in a single moment? Well, let's imagine what happens in your brain as you round a street corner and are confronted by a rhinoceros blocking your way.

First, your brain has to take in the raw sensations. The rhinoceros and every other object in the field of view reflects light of various intensities and wavelengths. This light hits the back of your eyeballs, where it is picked up by a layer of pigmented nerve cells, the retina. The scene is then relayed to the occipital lobe at the back of your head. A map of what you are seeing is projected (upside down) onto the primary visual cortex – a palm-sized, 1/12-in (2-millimeter) thick sheet of about half a billion densely interconnected neurons.

surprise, surprise
Every visual scene is packed with information. And yet we can decode it at a glance. We may never have seen this particular street before, but we can quickly make sense of it. And quickly spot what shouldn't be there.

optic path

Visual information from the eyes is transmitted to the back of the brain for processing. When it reaches V1, the primary visual cortex, it is also "projected" upside down.

Are you aware of anything yet? No, even though there has been some initial sharpening and tidying up of the picture. The primary visual cortex, or V1, is simply the first staging post. The serious analysis is to follow.

The human cortex has about 30 modules devoted to extracting the detail of visual experiences. Physically, the cortex is a single continuous sheet. Logically, however, it is broken into a mosaic of processing areas, with each patch of processing connected to the next patch, which builds to form an ever-rising stack of activity.

So you round the corner. A pattern of information is splashed across the primary visual cortex. A hierarchy of visual modules then starts to make sense of this pattern. The very next level, V2, is important for highlighting the boundaries and contours of what you are seeing. Like a heavy black marker, it draws around edges so that the shape of the object stands out from the shapes of the

cortex areas

The cortex of the brain has about 30 visual processing areas, fingernail-sized regions of gray matter that analyze what the eyes see. V1 is the primary mapping area. V2 is needed for boundaries, V3 for form and depth, V4 for color, and V5 for motion.

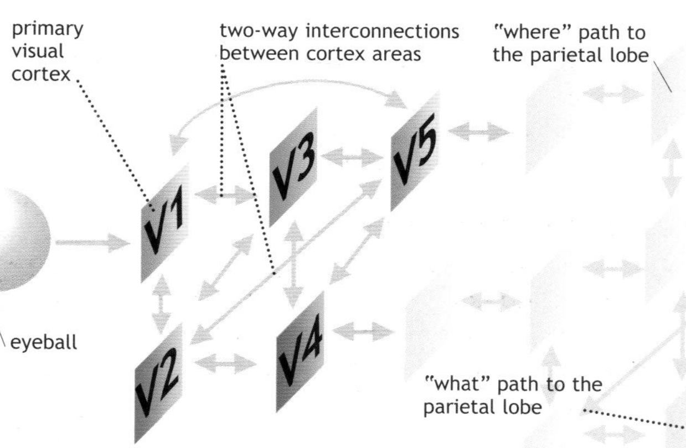

background. Next there is V3, which takes the analysis of shape and perspective a step further, while V4 takes the wavelength information and turns it into an experience of color. As the analysis continues, each level feeds the activity of the next to extract increasingly precise details. The brain cannot swallow the visual scene at one time, so it pulls it to pieces, distilling the information progressively. Sight is actually a bundle of senses – for color, motion, form, and so on – all "flying in formation."

branching pathways

Rather early on in the visual process, a fork appears in the visual hierarchy. One branch heads up the back of the skull toward the parietal lobes, focusing on "where is it?" questions to do with motion and location. The neurons are tuned to pick up shifts of position and also to subtract our own motion from the emerging picture so we can tell whether we are moving toward an object or the object is moving toward us. At the peak of this brain path emerges a general feel for the space about us – how near or distant are the many elements of the visual field.

The other branch of analysis runs down the brain toward the temporal lobes, and this focuses on "what is it?" questions. Driven by a gathering tide of sensory details – a distinctive silhouette, a horny gray hide, an impression of great bulk – these areas can put an identity to the visual object. Using the memories etched into their synaptic connections, they will splutter the answer "rhinoceros!"

key questions

The two crucial questions the brain must ask to make sense of any event are "what is it?" and "where is it?" Different processing regions handle these questions individually.

what
?
where

mapping the world

Yet at this stage the visual pathway is still busy mapping and recognizing everything in the field of view. As you turn a corner, you must take in the whole scene before you can zero in on any of its parts. Other neurons will be energetically signaling people, houses, cars, trees. Somehow the "rhinoceros" neurons need to make themselves heard over the general hubbub of activity.

So far we have been talking about the back half of the brain, the sensory cortex. Each of the senses – touch, taste, hearing, balance, and sight – has its own corner of this sheet. And each follows the same hierarchical logic. With hearing, for example, a pattern of raw frequencies is mapped onto the primary auditory cortex. Then further stages of processing identify and locate the sounds. The next phase of brain activity must draw in the frontal cortex.

memory snapshots

For the sake of completeness, we should mention that as the sensory hierarchies reach a peak, they also begin to converge. In special "multi-modal" areas such as the entorhinal cortex and especially the hippocampus, neurons start to fire in response to combinations of noises, visions, smells, and other sensations. The many strands of sensory activity are tied together. This does not mean the hippocampus suddenly lights up with miniature sensory images. But a pattern of activity forms that "points back" to a stack of other neural patterns. The hippocampus reflects the fact that certain details exist, spread across the rest of the sensory cortex circuitry.

basal ganglia

hippocampus

inner brain
Tucked deep inside the bulging cerebral hemispheres is the hippocampus, crucial to memory. The basal ganglia, essential for habit learning, is buried even deeper.

Among other things, this allows the hippocampus to be a memory organ. A snapshot of neural activity at the level of the hippocampus can be used later to cause firing to follow a reverse path back down the sensory hierarchy, recreating the feel of the original experience. Memory is a general property of brains, as every synaptic connection is shaped by experience. But the hippocampus is positioned to capture specific memories, to trap a neural template of a particular sensory state, and then to use that template hours or even years later to reconstruct that moment.

interpretation

Anyway, getting back to our rhinoceros example, first the sensory half of the brain maps everything – all the details. Next, the frontal lobes need to kick in. The sensory picture has to be focused by a frontal act of attention that then leaves the brain in a state of intention – knowing both what is happening and what it wants to do about it.

> **"Our mental machinery knows everything that is going on around us but discards most of it as unimportant before consciousness is reached."**
>
> Ulric Neisser, psychologist (1976)

The very highest levels of the brain are the dozen or so processing areas that make up the prefrontal lobes. These areas are not interested in what is routine about any moment, just what is significant and worthy of closer examination. It is when news of a rhinoceros-shaped patch of visual activity hits this part of the brain that a sharp conscious experience can start to dawn.

Acting in concert with the arousal and emotion centers of the lower brain, with which they have intimate connections, the prefrontal lobes organize a state of attention. The paying of attention has the effect of actually defining the sensory experience. The sensory cortex begins by responding to the full panorama – the rhinoceros –

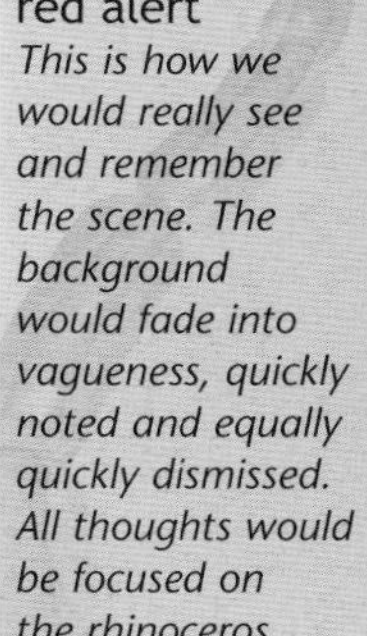

red alert
This is how we would really see and remember the scene. The background would fade into vagueness, quickly noted and equally quickly dismissed. All thoughts would be focused on the rhinoceros.

key points

- Sensory cortex maps sights, sounds, and smells.
- What? and Where? questions are answered.
- Hippocampus captures snapshot memory.
- Prefrontal lobes attend and respond.
- Emotions and associations are evoked.
- Physical response is coordinated.

but also everything else. Attention goes back over this to create contrast. It turns up the volume of the neurons representing the rhinoceros, pushing irrelevant details such as the presence of people, houses, cars, and trees into the dim fringes of that moment's awareness.

This is one of the advantages of the brain being an organic network of connections rather than a strict input-output device. The "output" areas can reach back down to alter their own inputs. Any input actually starts as more of a suggestion – "Hey, I might be important." Many inputs are analyzed, and eventually there is feedback to say "yes, you were the bit that mattered". The brain's pathways actually evolve a response.

mounting a response

A second obvious effect of paying attention to a particular aspect of a moment is that it unleashes a flood of thoughts, emotions, and associations. The whole of the brain is prompted to respond to the focal event with

cortex facts

The cerebral cortex makes up three-quarters of the total volume of the human brain. Spread out, the human cortex would cover about four sheets of 8½ x 11 paper. By comparison, a chimp's would cover a single sheet, and the cortex of a rat would cover barely a postage stamp. The frontal lobes – which include both prefrontal and motor output areas – account for 41 percent of the human cortex. The temporal lobes account for 22 percent; the parietal lobes 19 percent; and the occipital lobes 18 percent.

rat cortex

chimpanzee cortex

human cortex

whatever resources are available. So noting the sight of a rhinoceros will bring any stored mental associations to the surface. Even as we are recognizing the animal, relevant thoughts will be bubbling up in us. Has it seen us? Is there a zoo nearby? Should we freeze or run? Could this be a practical joke or maybe even a dream? Our arousal centers will be making decisions about whether to relax or whether to pump up the body for action. Our motor areas, behind the prefrontal regions on the frontal cortex sheet, will be gearing up to execute any intentions beginning to form. And even our language areas may start to organize a suitable (or suitably strangled) vocal response.

The brain thus takes in the whole of each moment and extracts, and responds to, its core aspect. A lamp-post or parked car may have been clearly visible alongside the rhino. But it is unlikely we would have any thoughts or feelings about them. The brain's mission is to find out what matters most and then to respond to that in as full and as coherent a way as possible.

So now lift your eyes again. Isn't it amazing that so much mapping and analyzing must be taking place for you to have any kind of sensory panorama? It may appear to happen almost instantaneously, but in fact it doesn't. It takes time, and that's another problem that brains have to overcome.

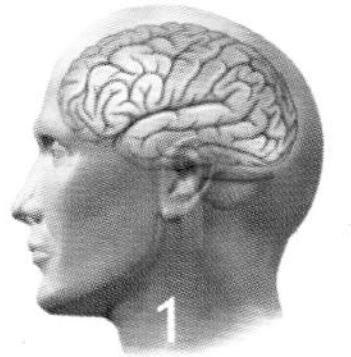

first, the mapping of raw sensations in the visual cortex

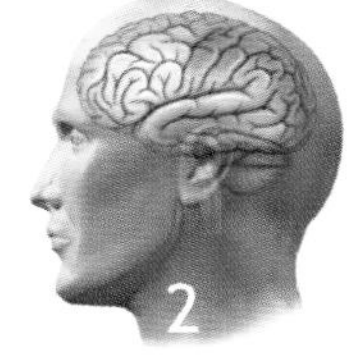

"what?" and "where?" paths recognize objects

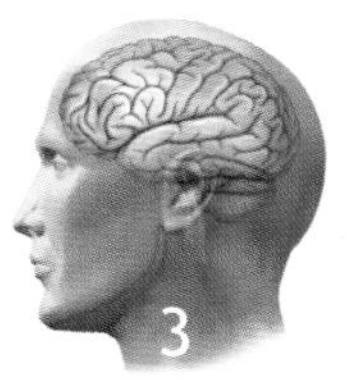

prefrontal lobes react and begin the response

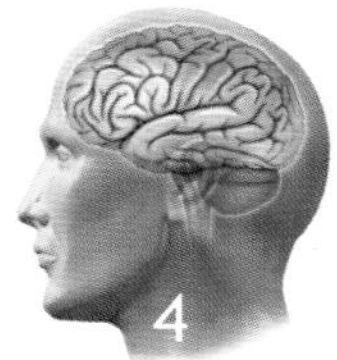

motor, planning, and emotion areas flood with thoughts

stages of processing

Sensations have to be processed. And this happens in a series of stages that extracts the details and digests the meaning. Only what is significant during a moment – such as seeing a rhinoceros – gets the full treatment as shown here.

bridging the gap

Have you ever played tennis? Good players must have incredible reflexes. In the professional game, the serve is banged down at 125 or even 140 miles per hour (200–225 kph). The ball is across the net and at you in about a third of a second. This seems barely time for an ordinary person even to begin a turn toward the forehand or backhand side. Yet top players can see the ball well enough to swoop and swing their rackets through the correct spot with split-second accuracy. At these speeds, just a few milliseconds (thousandths of a second) early or late and the result is an air shot – a missed ball.

time lapses

So you would think that the brain and nervous system must work lightning fast. But they don't. Even conducting signals at several hundred miles per hour, nerves need considerable time to turn input into output. It takes at least 20 milliseconds for messages to travel the length of the body. Because the eye takes a little time to register changes in light, visual signals take more like 50 to 100 milliseconds to reach the brain. Once inside the brain, many further connections are needed to transform the raw signals into some kind of mental response. Add up all these delays and it should be physically impossible to watch a ball right onto the strings.

Indeed, laboratory experiments show just how long it takes to integrate new information. When people are asked to tap a button as soon as a light flashes, it takes them 200 milliseconds – fully a fifth of a second. About

quick reactions
A tennis serve should be unreturnable. Nerve messages don't seem to travel fast enough. But habit and anticipation can help our brains beat time.

120 milliseconds are needed to register the fact that the light has flashed and another 80 milliseconds to get their finger to move. And this reaction time is for a simple, unthinking task. For any task that demands careful attention, such as the conscious juggling of several alternative responses, the response lag is nearer to half a second. Even more astonishingly, the greatest athletes are just as slow on raw reaction-time tests. Their brains work no faster. They cannot see what is happening in the world any quicker than the rest of us.

false starts
The human brain cannot react instantly – and that's official. At the Olympics, pressure sensors in the starting blocks automatically signal a false start if a sprinter pushes off within 120 milliseconds of the bang from the starter's pistol. The fittest, fastest, most highly trained athletes in the world still need time to know they have heard the gun and to tell their legs to start moving.

Brain processing does take time, too much time, it seems, for anyone ever to manage a game of tennis – or drive a car, or flip pancakes, for that matter.

processing shortcuts

Brains do ease the problem by always taking as little time as possible. To go all the way up and down the brain's processing hierarchy to form a fully conscious, attention-level mental response takes about half a second. Bringing hundreds of cortex areas into a state of coordinated response demands a lot of work. But the brain can learn to short-circuit this full-scale response and instead react out of habit, cutting the processing time from 500 milliseconds to "just" 200 milliseconds or so.

There are lower brain structures specialized for making this happen. A cluster of nerve centers, the basal ganglia, nestles inside the cerebral hemispheres, quietly noting the patterns of attention and decision making taking shape on the cortex sheet above. By watching, the basal ganglia begin to see which sensory patterns later produce a particular response. They can then literally short-circuit the production of that output state. As soon as the right kinds of sensations start to come in, the basal ganglia can trigger the same response in an immediate, unthinking way. The job can be done as if the higher brain had patiently considered its response.

cerebellum

The basal ganglia are not the only part of the brain watching and learning from the rather laborious activities of the cortex. The cerebellum – or "little brain" – is a fluted structure hanging off the back of the brain stem. This is a specialist organ for the fine-grain anticipation and timing of movements. The cerebellum accounts for just a tenth of the brain's volume, but it contains well over half its neurons – it has many more brain cells than the "smart" cortex. Neurons in the cerebellum also make 20 times as many synaptic connections – nearly 200,000 compared to about 10,000 for an average cortex neuron. However, the cerebellum uses rather simple processing loops. All the hardware is there to perfect the timing of actions, rather than forming a complex conscious picture. That is the cortex's job, with its branching connections and sensory maps.

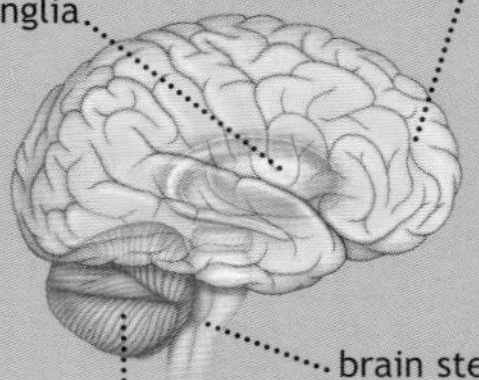

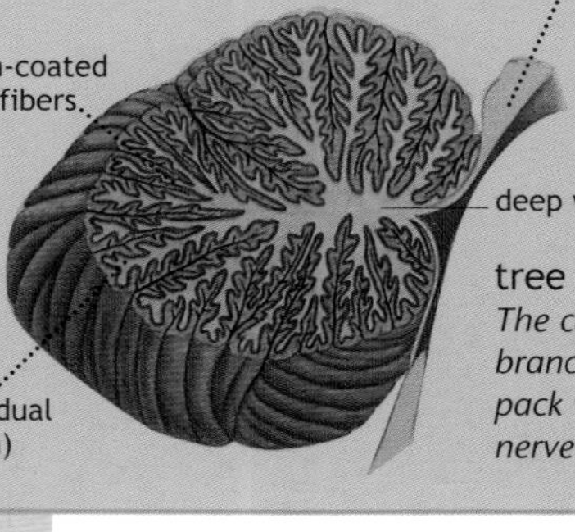

tree of life
The cerebellum has a distinctive branching structure, enabling it to pack vast numbers of myelin-coated nerve cells into a small volume.

This is a clever time-saving trick that works once a brain has experienced the same situation on sufficient occasions for it to become wired in as a habit – as a fixed-action pattern or automatism. But such a shortcut only reduces the response lag from a period of half a second to a fifth of a second. There still remains a yawning gap that needs to be explained, given that even walking down a flight of steps or pouring a kettle of boiling water are tasks that demand millisecond-level precision.

brain anticipation

The answer is simple. Brains anticipate. They get good at guessing. The brain can more or less assume that each new moment is going to be pretty much like the last one. Even when there is some wrenching surprise, like a car crash, we will still find ourselves inhabiting the same body. The same laws of physics still apply.

"Almost everything we do, we do better unconsciously than consciously. In first learning a new skill we fumble, feel uncertain, and are conscious of many details of the action."

Bernard Baars, psychologist (1988)

So the brain predicts that whatever it knew to be true about the world a moment ago will continue to be true. Against this simple prediction, it can then generate more specific states of expectancy. This is where attentive-level processing starts to count. It arrives too late to see the world as it happens, but it focuses our view of what has happened so that we are armed with definite expectations about what might happen next.

If we actually came across a rhino in the street, the surprise would be such that we would notice a gap in our reactions. We would feel a half second of

tricky footwork
Ever come down the stairs on a dark night and taken a step too many, reaching for one at the bottom that wasn't there? When expectations are confounded, our reliance on mental anticipation can be painfully brought home.

confusion as we struggled to make sense of the scene. But the very act of focusing on the unexpected existence of the rhinoceros would in turn generate a state of anticipation that would pave our way into whatever followed. We would be primed for the rhino to turn and fix us with its beady eye. Or perhaps mope off down the street. In some real sense, we will be living in the future, already conscious of what is about to happen. If our predictions are good enough, we can even begin to act without waiting for futher sensory input.

predator and prey
In the evolutionary arms race, all animals have nerves that work at the same speed. So IQ has to make the difference. Other little tricks, such as silent wings, razor claws, and phenomenal eyesight, will certainly help.

a fast response

This is certainly how athletes cope. They act on predictions. Experiments show that professional tennis players can guess the direction of a serve simply from watching their opponent's windup. Just the first split instant of the ball's flight is enough to extrapolate its full trajectory. Yet if the ball does do something unpredictable less than 200 milliseconds away from a player, such as take a bad bounce, then there is no time for adjustment. The player will simply swing on the basis of a wrong prediction and miss. Professionals read the game. They rely on anticipation. We ordinary mortals do exactly the same to cope with the millisecond-level tolerances involved in skills such as changing gears on a car or drinking from a cup without bashing our teeth.

Processing lags are a real problem for the evolutionary design of brains. If you are an antelope being chased by a leopard, or an owl swooping on a mouse, then you will want to be instantly aware of the world. The physiology of nerve cell transmission means this is never going to happen. But brains not only got around this, they have even gone one better, by learning to work ahead of the game.

the moment of awareness

The brain's job is to optimize behavior. It must understand the needs of the body and then recognize when the world offers opportunities to meet these needs. It does this by adapting its structure – the patterns of linkage that transform sensory input into motor output – across an ever-tightening series of timescales.

First, there is the genetic timescale. With each generation, the genes make a prediction about what kind of nervous system will do the job. Then Darwinian selection acts on this prediction, providing the feedback that leaves the genome better focused for its next round of circuit-building.

Second, there is the developmental timescale. Once brains swelled to a certain size, genes could only code for a

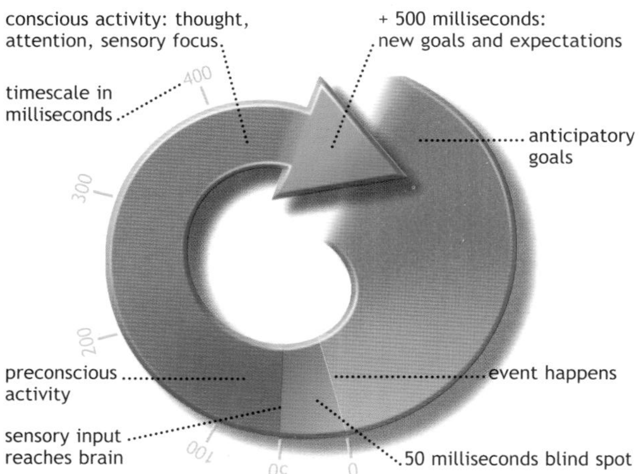

cycle of awareness

The brain begins with expectations and intentions. Then the moment happens. A split second later, the brain begins to map the sensations, extracting the meaning of the moment. The conscious results then become the context for the next moment.

William James (1842–1910), the founder of American psychology, popularized the idea of consciousness as a stream of mental events. He realized that despite our feeling of being a conscious self directing the show, in fact there is no such psychic core. Rather, a person is a collection of memories and habits that shapes the moment-to-moment flow of the mind. Selfhood is an accumulation of history.

rather general state of circuitry. Individual organisms had to build their own brain pathways, learning exactly how to sense and respond.

Finally, like the eye of a storm, we get down to the moment-to-moment flurry of adaptation that produces sharp awareness – the tightly specific pattern of linkage we experience as a focused state of knowing and intending.

getting oriented

We begin with a set of expectations. Everything – from the thoughts and experiences of the last few seconds to the dim memories of childhood – helps to get us oriented. Then the moment starts to happen. For about a twentieth of a second, we continue blindly to ride our wave of anticipations. But soon updated sensory impressions form. After a fifth of a second, the patterns are clear enough for the brain to react out of habit.

If unthinking habits have automatically dealt with as much as possible, then by definition anything that has not been handled by this stage must be novel, difficult, or in some other way unusual. It is therefore worthy of global, attentive-level, thoughtful processing. We may have bumped into a rhinoceros, crunched a gear change, stumbled on a step, or slopped a cup of coffee. Now the brain must adapt

frantic activity

Knock over your drink and you will flood with thoughts, actions, and feelings. Your heart lurches, one hand goes for the glass, the other to rescue your meal from the spreading puddle. Curses and apologies spring to your lips. A lot can be packed into half a second.

its neural state on the fly, evolving a pattern of linkage that best copes with this novel set of circumstances.

Feedback from the highest levels of the brain sharpens any sensory impressions, making them stand out. Attention also drives a brain-wide search for a meaningful response. Any memory association, emotion, or action pattern that might help is evoked. After about half a second of frantic activity, we are successfully reoriented. We know what has just happened and what we might want to do about it. The cycle of processing is complete.

the flow of consciousness

One final puzzle is that our processing-cycle story ends up making consciousness sound like a rather clunky production. Yet consciousness runs smoothly. Our awareness feels seamless for several reasons.

First, there is never an actual break in the stream of brain activity. The brain is always in a general state of knowing, a general connective balance. All the cycle of processing does is fleetingly tighten the focus of its circuits in certain directions. So moments of attention are like faster eddies marking an already flowing river.

In fact, most moments do not actually demand a radical change in the state of the brain. Anticipations smooth our path, making changes ahead of time. Habits handle much else without troubling the brain significantly. And, of course, most moments of our lives are actually rather routine, so any attention-level response will be muted.

Every waking moment may contain the potential for a seismic realignment of the brain's circuitry – a startled jolt. But there does not have to be such a jolt. The brain is quite happy rolling along like a sluggish stream, relying on anticipations and habits to assimilate the events of life as smoothly as possible.

rustling circuits

Shut your eyes. What do you see? A shimmer of light and movement. Your restless neural circuits are firing, even in the absence of visual input, to produce a play of color and shape. Without sensory input for a short period, brain cells tick over in their memory state of firing. Even this low-level rustle will produce a conscious experience.

the human mind

Clearly, there is a big difference between the minds of humans and other animals. An animal has awareness, but it is not clear that even the smartest beasts have self-awareness. Animals have intelligence, but they do not seem to plan, remember, and imagine in quite the same way we do. It is hard to put a finger on the exact nature of the difference, and still harder to be sure of the reasons. It could be just a matter of quantity – humans simply having bigger brains. Or it could be a difference of kind. We may have evolved an entirely new type of brain. The problem is that the human brain is not that huge – animals like whales and elephants actually have bigger brains. Nor is it visibly different in structure. A chimpanzee brain looks virtually the same. Perhaps it is a case of same hardware, new software? Speech was a *Homo sapiens* invention and language may help explain our suddenly expanded mental abilities. One certainty is that the mental transformation of *Homo sapiens* happened incredibly quickly.

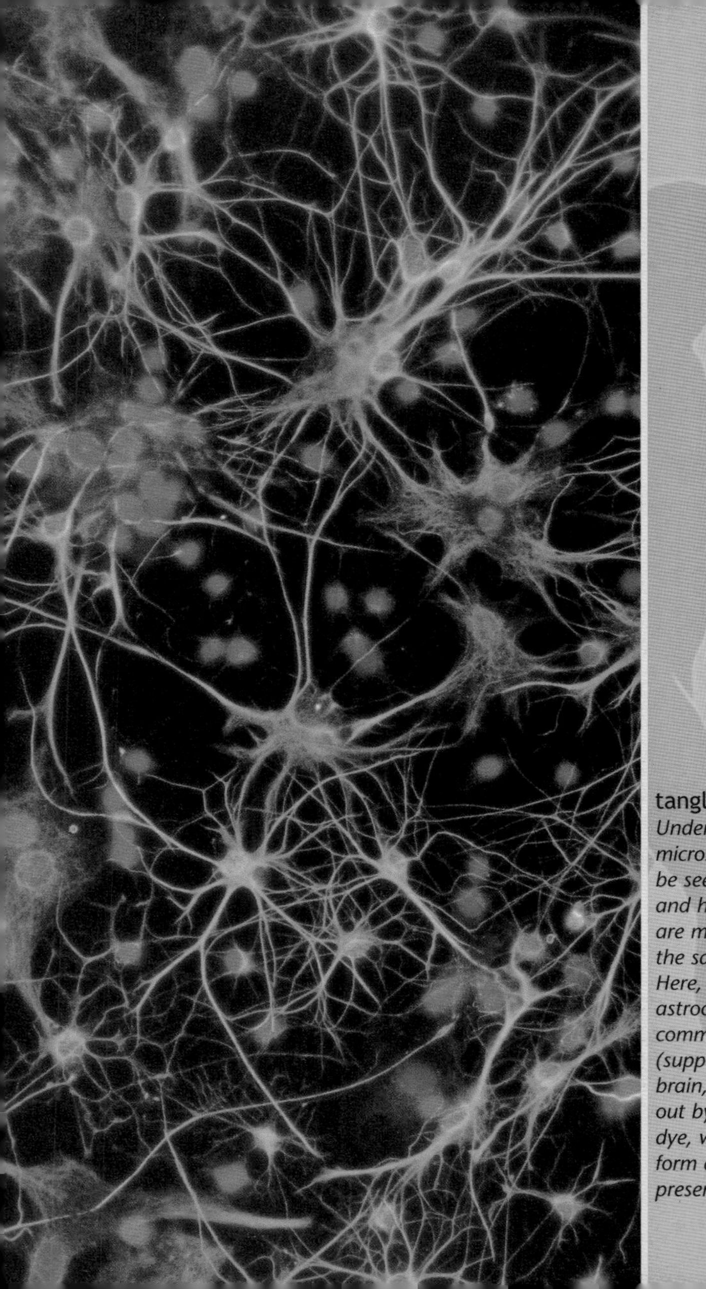

tangled web

Under the microscope, it can be seen that animal and human brains are made of exactly the same stuff. Here, star-shaped astrocytes, the commonest glial (support) cells in the brain, are picked out by fluorescent dye, while neurons form a ghostly presence.

the human riddle

Humans and chimpanzees are 98.6 percent identical in terms of their genes. Gorillas and orangutans are only slightly more different from us at 98.4 percent. Yet clearly, the mental gulf between humans and other apes is immense. The reason must be due to more than the simple fact that the human brain is actually four times larger than that of an ape. After all, elephants and whales have even bigger brains – their brains are three to five times larger than a human brain.

how smart?
Chimpanzees use twigs to fish termites out of nests, stones as hammers to crack tough nuts. They even use leaves to wipe their bottoms! But their intelligence has strict limits.

expanding consciousness

So far we have been telling a tale of evolutionary continuity. Taking the holistic route, we have been looking at how brains are the expression of a purpose. Brains exist to make decisions and they do this by following an adaptive trajectory. Over generations, over lifetimes, and finally over split-seconds, they zero in to form neural patterns that "know" the world. Thinking about brains in this light, we can see that the brainless activity of a tumbling bacterium is not that different in principle from the swift mental accommodations achieved by the large, complexly organized brains of vertebrates such as ourselves. But, with humans, we are forced to admit that consciousness takes on a different quality. And a glance at our evolutionary history reveals just how abruptly this change actually happened.

Some four and a half million years ago, a branch of apes – the hominids – learned to walk on two legs. There were many different hominid species that roamed the

human evolution

Human history involves much guesswork because few fossils remain. But it seems a small species of ape first walked upright about 4.4 million years ago. By about 3 million years ago, there were several hominid species in Africa, including "Lucy," a three-foot tall *Australopithecus afarensis*. By about 2 million years ago, the species had divided into two family lines – the heavy jawed vegetarian *Paranthropus*, and the lighter-boned *Homo*. Tool-using, meat-eating *Homo erectus* emerged 1.8 million years ago soon supplanting every other hominid species. *Erectus* expanded out of Africa to colonize Europe and Asia. During the next million years, *erectus* spawned several sub-species. In Europe, Neanderthals flourished from 300,000 to 30,000 years ago. Eventually another offshoot, *Homo sapiens*, emerged in Africa some 100,000 years ago, spreading even more quickly across the globe to become the new sole survivor of a complex family tree.

1300-1600 cc

400-550 cc

850-1100 cc

neanderthal
(230,000-35,000 years old)

australopithecus
(2-1.5 million years old)

homo erectus
(1.8-0.5 million years old)

1100-1400 cc

early human
(approx 120,000 years old)

bigger brains

These skulls reveal a steady increase in brain size (see labels above), and also a radical change in lifestyle. The jaw gets lighter, the teeth smaller, as hunting and cooking transform the diet. With humans, there is also a flattening of the face to allow for articulate speech.

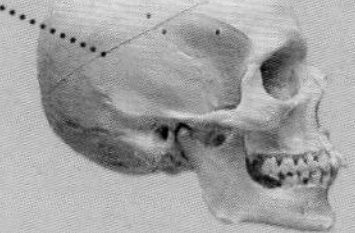

1200-1500 cc

modern human skull
(from 40,000 years)

African landscape. By about 1.8 million years ago, one of these had evolved into *Homo erectus*, a strapping six-footer with a brain three times the size of a chimp. *Erectus* was a hunter, a stone-tool user, and a fire user. So here was an apeman with a pretty smart mind.

Yet, *erectus* achieved strangely little following an initial burst of technological ingenuity. Having learned one design of axe, *erectus* went on making almost exactly the same tool for the next million or so years. There was a puzzling and intriguing lack of imagination.

the evolutionary explosion

But then *Homo sapiens* appeared on the scene some 100,000 years ago. Overnight the picture changes. Suddenly we are on a technological fast track. There is an explosion in the variety of tools being made and used for a variety of purposes. There are fish hooks, harpoons, bows and arrows, baskets, lamps, flint lighters, sewing needles, and knives with handles. However, the truly telling difference is that we are a symbolic species.

From about 40,000 years ago, we were daubing ourselves with paint, carving figurines and beads, even decorating our caves with hunting scenes. Our dead were buried with ritual. This archaeological evidence leaves no doubt that we had become fully modern, conscious of ourselves, our pasts, and our futures. We could think innovatively and had enough imagination to be ruled by superstitions and fears. So what could possibly have wrought such an intellectual transformation?

cave art
Nothing could be clearer evidence of a modern mind than the exquisite paintings found deep in caves in France and Spain, which were created from 34,000 years ago onward.

unlocking the brain

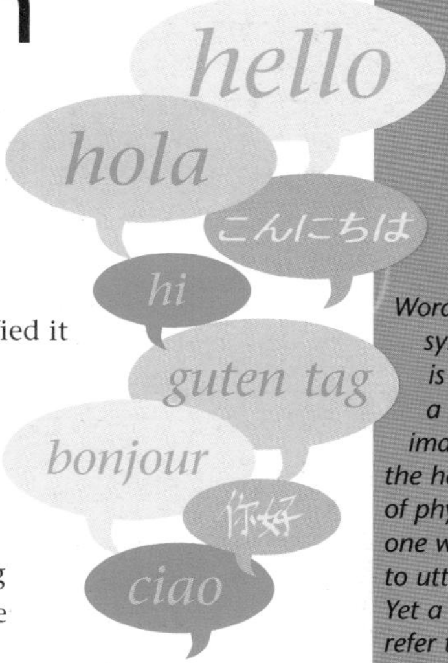

Let's say right away that this mental revolution must have been the result of language – the development of grammatical, articulate, referential speech.

While many theorists are satisfied it was language that unlocked the human brain, there are other suggestions. One is that our hominid ancestors evolved brain modules for a multitude of isolated skills such as tool making and social intelligence, then some freak brain mutation suddenly connected all the modules to create general intelligence. Yet there is little evidence the human brain was ever modular in the way the theory demands.

logical structures

When words are harnessed to the logical engine of a grammar – a set of rules for combining a stock of verbal tokens into coherent sentences – then speaking develops its own built-in momentum. All human languages share the same basic subject-verb-object structure (although, of course, different languages arrange these components in different orders). To speak a sentence, grammar forces you to tell a complete tale of who did what to whom. You have to take all the complexities of a real-life situation and shoe-horn them into a simple linear narrative.

symbolic signals

Words are symbols. A word is no more than a puff of air, an imagined noise in the head. In terms of physical effort, one word is as easy to utter as another. Yet a sound may refer to anything from a mundane object – such as a banana, a table, a rock – to abstract concepts such as valor, destiny, chaos, or the Universe. Words are capable of making the most intangible thoughts accessible.

Even a simple descriptive sentence such as "the cat slept" depends on this hidden logical structure. Here, the cat happens to be both the doer and the done to. We might instead have said "sleep overtook the cat" – especially if we wanted to imply some level of resistance on the part of the animal. Words and grammar allow for the finest shades of meaning. But the rules of sentence-making always force you to make a logical assertion. You have to choose from many possibilities to state something specific.

Then, of course, having made a precise assertion, either you or your audience will immediately want to respond. Every sentence provokes thoughts about what might be right or wrong with it, how it could be corrected, pursued, expanded. Words always lead to more words. Like giving a worm a head-end to its body, the sequential logic of the speech act creates direction. Thoughts expressed in sentences cannot help but be going somewhere.

recap

The **animal brain** is conscious. The human brain has something extra. It is self-conscious. We are a "self" – aware that we are aware. Speech was the mental tool that enabled us to turn in upon ourselves.

living in the moment

Why is this so important to the human story? It is because the minds of animals are trapped in the present tense. The very nature of the brain locks them into thoughts about what is happening right here, right now. Language allowed the mind of *Homo sapiens* to break free.

The brain evolved as a mechanism to make decisions. The nervous system encodes patterns that link inputs to outputs. As animals developed more and more elaborate brains, they came to understand more and more about the opportunities and threats contained in each passing instant. The full weight of the brain's history, all its remembered

trapped in the present

Famously, goldfish do not remember anything for more than about ten seconds. But that also means they do remember things – such as which way to swim for food or how to avoid a shock – for about that long.

experiences, remembered intentions, and remembered expectations, could be applied to the processing of a moment. The result was a sharply intelligent awareness of the world. An animal such as a chimpanzee understands both what is going on at any given moment and what it wants to do about it. It can imagine futures that arise directly out of the present, but not futures unconnected with some present experience.

However, this relentless pressure to extract meaning from the passing instant leaves an animal with its nose pressed hard against the windshield of life. An animal does not lack for consciousness, but it is a viewpoint that simply looks into what is currently happening. Greater brain power merely thrusts an animal even further into an intelligent appreciation of the threats and potentials of the world that surrounds it. A slug does not know much about a moment. A goldfish or frog has a hazy awareness of the moment. But a dog, dolphin, or chimpanzee is quiveringly alert and emotionally responsive to the smallest details of a moment. And yet even the brainiest animals are still being driven by the demands of the here and now. What brains needed was some mechanism for stepping back, a way of redirecting their phenomenal responding powers and turning them on problems or situations that were not immediately present.

living for the here and now

Dolphins have a remarkable ultrasonic sense. Like bats, they emit high-frequency clicks and can use the echoes to "see." Because sound penetrates, a dolphin can even see the child inside a pregnant woman swimmer. But despite being bright and playful, dolphins are still being driven by the here and now.

" We say a dog is afraid his master will beat him, but not he is afraid his master will beat him tomorrow. Why not? "

Ludwig Wittgensein, philosopher (1953)

expanding consciousness

Language was this mechanism. Speech – especially when internalized as an inner voice – gave us a new kind of software that could take our already conscious brains to places they had never been before. Words could transport us to imagined moments, or even imaginary viewpoints such as a perception of ourselves as we would be seen by others. Words created the mental distance by which we could become not just conscious, but self-conscious – able to contemplate the fact that we were a self with a history and a future, desires, and responsibilities. The biological machinery of attention and intelligent response was unchanged. But we could use the scaffolding of words to direct this attention inward and catch ourselves in the very act of responding.

" Let anyone try the experiment and he will see that we can as little think without words as we can breathe without lungs. "

Max Müller, linguist (1888)

inner speech

The use of inner speech to shape our thoughts becomes so habitual that – like changing gears on a car – we end up barely aware of what we are doing, concentrating instead on where we want to go. It is easier to appreciate the role played by self-addressed speech when eavesdropping on a three- or four-year-old child, whose play is guided by verbal monologue. This thinking aloud later becomes thinking silently. As adults, it is usually only when we face difficult problems that we revert to muttering to ourselves. By the same token, learning to read without pronouncing the words is an acquired skill.

grammatical brains

UGG!

From speech came symbolic thought. But how did *Homo sapiens* gain the sudden power to speak? In fact, words may have existed for millions of years before the invention of grammar.

Chimpanzees and dolphins can master a vocabulary of several hundred words, learning the names for objects and actions. So it is quite likely that *Homo erectus* could grunt single word commands such as "firewood!" A nod toward a guttering fire would then be enough to tell a junior member of the tribe exactly what was required. In a tribal setting, even a small vocabulary would have gone a long way.

But grammar – the ability to string words into sentences, or even inflect individual words to create different tenses, or make plurals and comparatives – is something else. A rule-handling brain was the key advance made by *Homo sapiens*.

language centers

It is well known that the adult human brain has language centers – two coin-sized areas of cortex specialized for word production and grammatical rules. The traditional view was that these language centers were wired in at the genetic level. But a better understanding of the cortex's development suggests that the only thing that can be

smart words

Ugg! In caveman talk, it could mean "charge", or it could mean "retreat." A word is a token, a mere puff of air. Yet words can have huge power to organize both our social and mental behavior.

key points

- Words may have existed a million years before the invention of grammar.
- The human brain has "language centers", but much of the brain is actually involved in speech production.
- Babies have instincts that help them pick up the rhythms of language quickly.

wired in by the genes is a propensity to learn certain kinds of perceptual and motor skills.

Recent brain scanning studies have shown that the structure of language actually soaks the brain. While the language centers may form the focus of activity, most of the brain, from the prefrontal cortex down to lowly structures like the cerebellum, light up in the effort of producing speech. This suggests that a rule-handling brain is the result of some general genetic tweaking rather than the evolution of a specific new grammar module.

If we compare a human brain to that of an ape, we can see a number of general differences that would pave the way for the easy learning of grammar. First, the human brain is four times bigger, so it starts with more cortex to devote to the task. Second, our brain is much slower to mature. As we have already explained, the human brain is still sprouting and pruning connections even into early adulthood. This allows a human child plenty of time to absorb the sounds and rhythms of language. A third general difference is the lateralization of the human brain – its famous division into left- and right-hand halves with distinct processing styles.

hand signs
The deaf think in sign language. Just as we hear the constant chatter of a small voice in our head, the deaf feel imaginary hands signing thoughts in the privacy of their minds.

language in the brain
These PET scans show, in red-orange tints, the areas of the cerebral left hemisphere that are active while understanding language. The left-hand scan shows activity associated with deciphering the meaning of words. The right-hand scan shows activity associated with understanding sentences.

tuning in to speech

Babies are born with instincts that will help them quickly to tune into the rhythms of speech and grammar. They are very sensitive to the direction of another person's gaze – which helps them know what their mother is saying when she gushes "Look at that funny dog!" They naturally experiment with gurgling and babbling noises that tune up their vocal cords. They are also instinctive turn-takers, and will alternate between bursts of babbling and rapt attention to a parent's cooing. This behavior helps them to learn the to and fro pattern of conversational speech. So while language may not be hardwired into the infant brain, there are many subtle genetic tricks that ensure a baby starts to learn the right lessons from the day it is born.

left or right?

Popular mythology has it that our left cerebral hemisphere is the verbal, rational, pedantic side of our brain, while the right is the creative, visual, holistic side. There is a grain of truth in this characterization. But the actual difference is really one of attentional style, and the two sides always work together. The true story is that the left brain is good at taking a focused and sequential view of whatever thoughts are in mind, while the right is specialized for stepping back to consider the broader context. One side deals with the foreground of the task, while the other deals with the background.

So when it comes to speaking, the left brain acts in a selective way, picking out the individual words and applying the specific grammatical rules. In any test of language production, the left brain would seem to be

fact or fantasy?
The human brain is divided, and popular mythology has it that the left is logical, the right is imaginative. There is some truth in this characterization. But both sides of the brain are fully involved in any normal act of speech or thought.

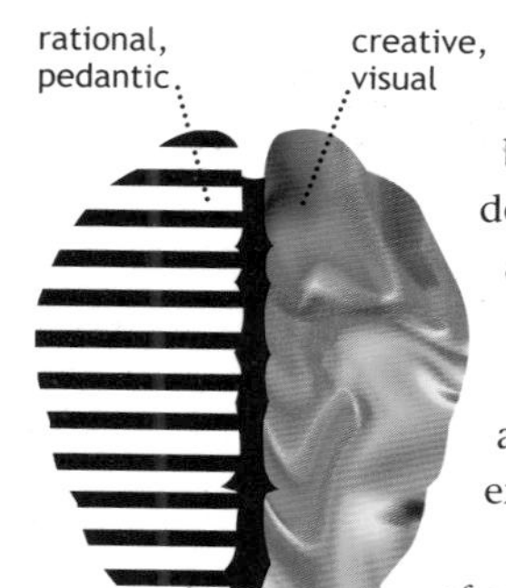

doing the critical work. But the right brain creates the background to a speech act, dealing with the more nebulous aspects, such as the emotional tone of what is being said, or the general sense of the sentence and where the thoughts it expresses might be leading.

There is evidence that the brains of apes are slightly lateralized in their processing. But early *Homo sapiens* may have developed a highly lateralized brain and that the handling of sequential grammatical rules suddenly became a very easy process as a result.

Tool-use may have set the ball rolling. Chipping away to shape a flint axe also requires holding a general aim in mind while executing a succession of steps to achieve that aim. So, perhaps *Homo erectus* evolved a divided brain to manipulate artifacts, then early humans exploited it to handle symbols. Whatever the answer, we know today we have minds that can formulate precisely what to say within a greater backdrop of what we mean to say.

The claim that animals are trapped in the present and that language sets humans free has been made by philosophers as varied as John Locke and Arthur Schopenhauer. In the modern era, the argument was most vigorously pursued by the great Belorussian psychologist, **Lev Vygotsky** (1896–1934), whose brilliant writings and careful experiments had little impact until recently. He died young, of tuberculosis, and most of his work was suppressed by the Soviet authorities. This seminal thinker was only rediscovered in the 1970s.

memories are made of this

Putting it all together, we can say that we began as a large-brained ape – intelligent but trapped in the press of the moment like any other animal. Our *erectus* ancestors probably used simple utterances to organize their social lives, giving them enough language to make them decent hunters and tool-makers, but not enough to spark an inner mental revolution. Then along came *Homo sapiens* with a brain flexible enough to absorb the complex rhythms of grammatical speech. It turned out that a tool initially developed for social communication between individuals could be used internally for organizing the thoughts of a self.

An animal mind cannot respond to a rhinoceros lurking around a street corner until it actually sees it, hears it, or smells it. Something about the current moment has to be giving good reason to be pondering such an unlikely event. But humans can use words to conjure up the mental image – a clear sensory anticipation of what such a moment might feel like. And combinations of words can be used to generate a sensory impression that is quite literally imaginative. Try thinking of a rhinoceros in a floral frock, chatting to a policeman, with a tennis racket in one hand and a poodle balanced on the other.

darwinism rules
We all know about Darwin (above) and genes. Now evolutionary theorists have coined a term for the cultural equivalent, the meme. Whereas a gene makes a protein, a meme is a socially useful idea transmitted by language or other symbolic means. Darwinian competition acts on memes to produce rapid cultural evolution.

memory aids

Words are important in recollection as well. All animals have memories. The brain itself is a memory surface, every connection a product of its history. Higher animals also have a hippocampus and the other specialist structures needed to take mental snapshots of individual moments. Seeing that we all have the same brain hardware, we should expect that they too can remember the look and feel of particular instants. But the difference is that an animal would need an external trigger to jog these specific memories back to the forefront of awareness.

A dog may remember a man who kicked it many years ago the instant it sees or scents the man. But does the same dog ever lie on a hearth rug having angry thoughts about old incidents? Or happy thoughts about a previous day's walk? Only humans have an inner voice driven by the relentless logic of a sentence-forming grammar that they use to probe their memory banks. The recollections themselves are mental images – sensory anticipations of what it would be like to be reliving some distant moment. But can we stir even fleeting states of recollection without going fishing for the details using the triggering power of words?

Well, can you? Irritatingly, you cannot read these words without responding. Sentences will struggle to form in your head. You will want to move your thoughts along to their next logical step. Words have a way of breeding words, always shifting your attention even further away from a response to the world immediately about you.

So now instead lift your eyes and look around. Here is the marvel of the human mind. You can switch between the inner and outer view. You can switch your focus at a command. And if it was not the mastery of speech that was the key to the abrupt freeing of the mind of *Homo sapiens*, then science is still scratching its head as to what else the answer could be.

dog days
It is hard to imagine your dog daydreaming about its far-off days as a puppy. Dogs can remember people, places, and smells, but it is not an autobiographical kind of recollection.

conclusion

There we have it. The human brain is the most complex object known in the Universe. Reductionist science wants to strip away as much of the complexity as possible. It aims to talk about consciousness in terms of simple brain mechanisms – the firing patterns of neurons, the hierarchical order of the cortex. Telling the tale in this way helps us to focus on critical details; an understanding of the parts does help in an understanding of the whole.

But reductionism must always be tempered by an appreciation of the bigger picture. A brain scooped out of a skull is just an oozing lump of protoplasm. A brain gains meaningful form by being plugged into a context, by having tasks it must strive to perform. Embedded in a body, a brain can light up with consciousness. It can feel the pressure of inputs and begin to adapt its circuitry in ways that express a specific state of output. Plug a language-enabled human brain into a cultural setting, a body of its fellow humans, and it can really start to go places, being shaped by social as well as biological needs.

Thus if your mind seems a busy space, filled with far more than appears credible for something generated by a mushy lump of circuitry, then you are right. You are the physical expression of so much more evolutionary, developmental, and social history than you ever realized. The brain is not a mechanism but a delicate organic entity in a constant frenzy of self creation. The richness of its pathways reflects the richness of our lives.

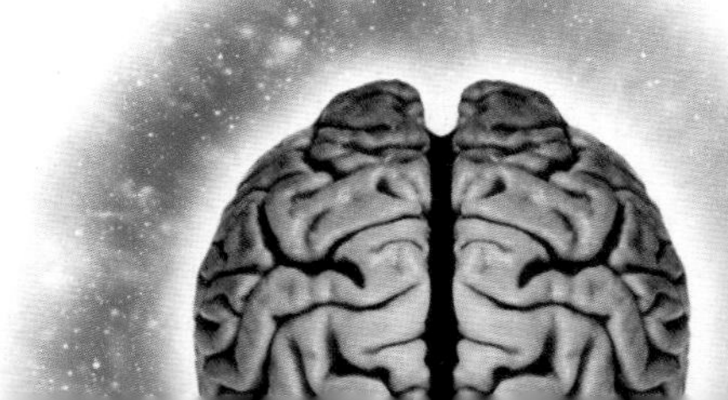

only connect
The complexity to be found inside a brain is simply a reflection of the complexity of its relations with the world into which it is born, and within which it has to live.

glossary

attention
The mental spotlight that is our highest level of processing. Attending both enhances our awareness of what is in focus and suppresses our awareness of what falls outside.

axon
Output arm of a neuron. Branches at the end to form usually a few hundred synaptic connections. Signals are transmitted in the form of spikes. Large, long-distance axons are wrapped in myelin.

basal ganglia
Network of nerve centers or ganglia hidden from sight at the base of the cerebral hemispheres. Crucial for controlling the timing of movements and the learning of habits.

behaviorism
School of psychology founded in US in early 20th century. Based on the belief that science cannot deal in mental states, only outward behavior. Behavior, in turn, had to be analyzed in terms of simple stimulus-response relationships.

brain stem
First thickening of the spinal cord as it enters the brain. Primitive region with important body-regulation and arousal functions.

central nervous system
The brain, spinal cord, and retina.

cerebellum
Large, fluted lobe that hangs off the back of the brainstem. Originally thought to be a movement coordination center, now known to "smooth" all forms of mental action by handling the fine-grain detail of timing and prediction.

cerebral hemispheres
General name for the cerebrum, the wrinkled mass of the higher brain. Formed by a pair of symmetrical lobes joined at the center by a thick band of nerves known as the corpus callosum. The surface of the cerebral hemispheres is the cortex, while the inside consists of white matter, fluid-filled chambers known as ventricles, and various lower-brain centers.

consciousness
Subjective awareness. Consciousness varies in grade from the bright, dominating focus of attention to the dimmer, fringe awareness associated with the performance of habits or the processing of sensations on the periphery of the moment.

cortex
The wrinkled surface of the cerebral hemispheres, from the Latin word for "bark" or "rind." Although this layer of grey matter is only a few millimeters deep, it contains a mass of white-matter connections and is divided into hundreds of processing zones.

dendrite
Input arm of a neuron. While neurons usually have a single axon, they have a bush of dendrites studded with thousands of synapses that are used for making connections with neighboring cells.

depolarization
Sudden change in the electrical potential across a cell membrane. Neurons build a state of polarization by pumping out positivelycharged ions, then depolarize by opening membrane pores that let the ions flood back in.

evolution
Simply defined as the survival of the fittest. Reductionists stress the element of blind competition, avoiding any suggestion that evolution might have purpose or direction. Holists, on the other hand, see fitness as depending on emergent properties such as mutual

benefit as well as naked competition, and are more comfortable with the idea that there may be a natural tendency toward greater complexity, autonomy, and even purposefulness.

frontal lobe

Front half of the cerebral hemispheres. Contains prefrontal cortex and motor cortex. Rear of the cerebral hemispheres map sensations and then frontal lobes organize the response.

gestalt psychology

School of psychology originating in Germany in early 20th century. It was founded on the idea that perceptions and thoughts come to us as already organized wholes rather than atomistic collections of sensory elements.

glial cells

Support cells in the brain. Important for health, growth, and regulation of neurons. They appear in many forms and outnumber neurons by 50 to one in some areas.

gray matter

Collective name for brain tissue, which consists of a tangled mass of neuron cell bodies, their dendrites and axons, and glial cells. Actually pinkish, but turns gray when pickled in formalin preservative.

habit

A learned mental action that can be performed without thought or conscious supervision. Also called a fixed-action pattern or automaticism.

hippocampus

Brain structure critical for forming memories. Formed by a roll of gray matter inside the temporal lobe. From Latin for "seahorse," its shape resembles a curved, tapering tail.

holism

Belief that the whole is more than just the sum of its parts. An alternative to reductionism, in that it seeks to deal with systems in their entirety.

hominid

Family name for the Hominidae, the branch of bipedal apes that arose about 4.5 million years ago and eventually gave rise to humans.

homo sapiens

Modern humans, from the Latin for "wise man." They are believed to have arisen in Africa about 100,000 years ago.

homo erectus

Believed to be direct ancestor of modern humans. Appeared about 1.8 million years ago in Africa and evolved into a number of species including eventually *Homo sapiens*.

ion

Electrically charged atom, usually a dissolved salt such as sodium, calcium, or potassium.

lateralization

Division of the cerebral hemispheres into two halves, each with different, yet complementary, processing styles.

memory

Changes in the connections of the brain that alter its future behavior. Memory comes in various forms – from habit-level learning to recollection, which is the mental reconstruction of past events.

millisecond

One thousandth of a second.

mind

A more general term than consciousness since it includes the unconscious, subconscious, and other such aspects of mental function.

myelin

Fatty, white-colored protein used to insulate axons, thus speeding up the transmission of signals in the brain. Produced by glial cells that wrap themselves around the axon.

neanderthal

Thickset "cousin" of *Homo sapiens*. Lived 230,000 to 35,000 years ago in Europe and parts of Asia. They were large-brained, but lacked sophisticated tools and symbolic culture.

neuron

A nerve cell in the brain. Neurons come in a great variety of shapes and sizes, but all operate on the same principle of summing input charges on their dendrites and cell body, and then firing a spike down an axon once a polarity threshold has been breached.

nuclei
In the brain, small masses of grey matter. From the Latin for "kernel."

neurotransmitter
Chemical messengers that carry the signal between neurons. The brain mainly uses small amino acid-based organic molecules such as glutamate, dopamine, and acetylcholine. Effects on the next cell can be inhibitory or excitatory.

occipital lobe
Lobe at back of cerebral hemispheres that deals mainly with vision.

parietal lobe
Cortex region that is sandwiched between the frontal and occipital lobes. Deals mainly with sensations of touch and motion, and with spatial relationships.

peripheral nervous system
Nerves of the body that connect the muscles and organs to the central nervous system.

prefrontal lobe
Region of the cortex immediately behind our forehead. Contains at least a dozen areas that are necessary for higher-level planning, control, and decision making.

primary visual cortex
First input area for visual signals from the eyes. Also known as V1. The largest single cortex region, most of it is actually buried from sight in a deep groove of the occipital lobe.

receptor
Large protein structure embedded in the membrane of a cell. It has binding sites that lock onto neuro-transmitters or other substances. Binding causes the receptor to change shape, either triggering processes on the inside of a cell or opening a pore in the cell membrane so ions can flow.

reductionism
An approach that explains complex systems by breaking them down into their constituent parts.

retina
Light-sensitive lining of the eyeball. The retina is actually a mix of pigmented receptor cells that detect the light and neurons, which are responsible for some visual processing.

self-consciousness
Consciousness of being a self. Animals have a sense of consciousness, an intelligent awareness of the world. But humans can turn this awareness around to take note of their own inner traffic of sensations, intentions, emotions, and expectations.

spike
Basic neural signal. Also called an action potential, a spike is a rolling wave of depolarization along the axon of a neuron. Information is not actually carried in individual spikes – which all look the same – but in the number, as well as the timing, of spike volleys. Neurons can fire as many as several thousand spikes each second.

spinal cord
Major nerve tract that carries signals to and from the brain. Runs through center of backbone, but ends at level of lower ribs. Has gray matter and can control reflexive actions such as maintaining postural balance.

synapse
Junction between two neurons. When membrane depolarization causes the release of neurotransmitter molecules on one side of the junction, this flows across, triggering receptors and causing changes in membrane polarity on the other side.

temporal lobe
Lobe of the cerebral hemisphere that curves forward toward the cheekbone. Deals with the sense of hearing, aspects of vision, and object recognition. Houses other key structures such as the hippocampus.

white matter
Collective name for the whitish mass of myelin-sheathed axons that fill the interior of the brain. White matter makes up nearly half the brain.

index

a

b

c

d

e

VA

Further reading

Going Inside – A Tour Around a Single Moment of Consciousness, John McCrone, Faber, 1999. ISBN: 0571173195

The Myth of Irrationality, John McCrone, Macmillan, 1993. ISBN: 033357284X

The Ape That Spoke, John McCrone, Macmillan, 1990. ISBN: 0333537920

Evolving Brains, John Allman, WH Freeman, 1999. ISBN: 0716750767

In The Theater of Consciousness, Bernard Baars, OUP, 1998. ISBN: 0195102657

Cognition and Reality, Ulric Neisser, WH Freeman, 1976. ISBN: 0716704781

Thought and Language, Lev Vygotsky, MIT Press, 1986. ISBN: 0262220296

The Web of Life, Fritjof Capra, Flamingo, 1997. ISBN: 0006547516

Hierarchy Theory, Howard Pattee (ed), George Braziller, 1973. ISBN: 080760674X

What's Going On In There? Lise Eliot, Allen Lane, 1999. ISBN: 0713992913

The Symbolic Species, Terrence Deacon, Allen Lane, 1997. ISBN: 0713991887

John McCrone's website – http://www.btinternet.com/~neuronaut

Author's acknowledgments
If there is a historical trajectory to the ideas outlined in this book, it runs through the writings of Thomas Hobbes, Wilhelm Wundt, William James, Lev Vygotsky, Aleksandr Luria, Wolfgang Köhler, Donald Hebb, Gilbert Ryle, Evgeny Sokolov, Ulric Neisser, Bernard Baars, Francisco Varela, Howard Pattee, Walter Freeman, Jeffrey Gray, David LaBerge, and Stephen Grossberg.

Additional picture research
Penni Bickle

Illustration
Richard Tibbitts, AntBits illustration

Index
Indexing Specialists, Hove

Picture credits
EMPICS: 43. **Novosti Picture Library**: 62(b). **Powerstock**: 19(b). **Science Museum**: 13. **Science Photo Library**: 11(t); 14(b); BSIP, ATL 32; Scott Camazine 9, 15(br), 21; Nancy Kedersha 51; Mehau Kulyk front cover, 1, 65; Dr Kari Lounatmaa 17(b); John Meyer, Custom Medical Stock Photo 15(bl); Hank Morgan 24(b); Sam Ogden 12(b); Alfred Pasieka 5, 48(t); Volker Steger 63; Wellcome Dept. of Cognitive Neurology 60(b).

Every effort has been made to trace the copyright holders.
The publisher apologizes for any unintentional omissions and would be pleased, in such cases, to place an acknowledgment in future editions of this book.